Electric Protective Devices

Electric Protective Devices

Protection with Energy Saving

Khalil Denno
Distinguished Professor of Electrical Engineering
New Jersey Institute of Technology, Newark, N.J.

—McGraw-Hill, Inc.
New York   San Francisco   Washington, D.C.   Auckland   Bogotá
Caracas   Lisbon   London   Madrid   Mexico City   Milan
Montreal   New Delhi   San Juan   Singapore
Sydney   Tokyo   Toronto

Library of Congress Cataloging-in-Publication Data

Denno, K.
    Electric protective devices : protection with energy saving /
Khalil Denno.
        p.   cm.
    Includes bibliographical references and index.
    ISBN 0-07-016422-3
    1. Electric power systems—Protection.   2. Electric power
transmission—Equipment and supplies—Protection.   I. Title.
TK1005.D433   1994
621.31'7—dc20   93-15651
                                                           CIP

1 2 3 4 5 6 7 8 9 0   DOC/DOC   9 9 8 7 6 5 4 3

ISBN 0-07-016422-3

*The sponsoring editor for this book was Larry G. Hager and the
production supervisor was Suzanne W. Babeuf. It was set in Century
Schoolbook by North Market Street Graphics.*

Printed and bound by R. R. Donnelley & Sons Company.

See Acknowledgments for credits to individual publications.

To Badia, Karem, Zayd, and Athra for their love, dedication, and loyalty.

Contents

Preface

This book presents a comprehensive analytical and design picture for electric protective devices in unique association with their role in energy saving and storage. Design composition of electrical devices involves: resistors, inductors, capacitors, saturable-reactors, saturable-resistors, arresters, and relays. Protective function of those devices will be linked with energy transport and storage, such as the hydrogen fuel-cell, the redox flow cell, the storage battery, and the magnetohydrodynamic generator, as well as the conventional induction generator.

Chapter 1 presents a principle review with specific examples for the applications of lumped elements of resistors (as surge dampers), inductors, and capacitors (as surge spreaders) at various traditional locations in the operational power system.

In Chap. 2, steady-state and transient properties for the new protective device, known as the *saturable-resistor* (*saturistor*), is presented in a comprehensive, analytical way. Protective function of the saturistor is centered on its behavior as an automatic surge damper and limiter with respect to increase in current surge and its frequency. Theoretical calculations coupled with experimental verifications for the induced fields inside the saturistor core, the variable time constant, and movement of the plane of magnetization, as well as eddy-current losses have been presented in detail.

In Chap. 3, protective function for the saturable-resistor has been studied in detail in applications involving the three-phase wound rotor induction machines, the single-phase induction machines, and the synchronous machines, as well as the conventional linear induction accelerator. In induction and synchronous machines, the saturistor provides damping for line current at starting and upon interruption and resumption of power as well as smoothing and optimization for the ratio of torque with respect to line current. For the linear induction accelerator, the saturistor through special control network provides the source of charging and accelerating beam current desired in heavy ion beam for nuclear fusion.

In Chap. 4, theoretical concepts have been presented for the storage of captured surge energies from high-voltage switching and that induced by lightning surge in ideal inductors, capacitors, and various modes of their combination.

In Chap. 5, process design and operational characteristics for energy storage of captured surge energies have been presented, including the storage battery, the redox flow cells, the magnetohydrodynamic generator, the electrogasdynamic generator, and the conventional induction generator, as well as the process of electrolysis and electroseparation for the release of H_2 and O_2.

In Chap. 6, device protection design, which includes the saturistor in various modes of connection with high-quality inductor and high dielectric capacitor, is presented. Calculations of output voltage or current for several protective devices due to a voltage or current input surge have been established.

In Chap. 7, systematic protective devices involving the saturable-resistor, ideal inductor, and capacitor, together with a nonlinear voltage and/or current arrester have been presented with principle steps for calculating its output voltage or current surge. The dual functions of protection and the saving of electromagnetic energy are stressed.

Chapter 8 presents useful summary for the subject of symmetrical components with clear outline for the modes of interconnection of sequence networks for all kinds of asymmetrical faults at no-load and at any loading.

Chapter 9 presents the subject matter of coordination and design of power polarity relays and distance relays in power system operation and control. The criteria of relay optimal selectivity, as well as design identification of protected region, are presented also.

Chapter 10 points to the realization that power electronics based on solid state technology—namely, the application of silicon thyristor—will be the cornerstone in devising the developing future bulk power system with more reliability and promptness in protection and control.

Each of the ten chapters includes a set of problems, with a special set of supplemental problems at the end of the book in the Appendix.

Khalil Denno

Acknowledgments

With special thanks to Ms. Chiung-Wen Haung for her help in performing the word processing of this manuscript. Also my thanks to Ms. Margaret Cummins, the editorial assistant at McGraw-Hill, Inc., for her prompt response and professional cooperation at all times.

The author would like to thank the following publications for release of data and diagrams for use and publication in this book:

©1970 Gordon and Breach Science Publishers, S.A. Reprinted with permission from *Induction Machines,* 2d edition by P. L. Alger. (Sections 2.1, 2.3, 3.1, and 3.5; Figs. 2.1, 2.3, and 2.12.)

©1963 IEEE. Reprinted with permission from P. L. Alger, G. Angst, and W. M. Schweder, "Saturistors and Low Starting Current Induction Motors," *IEEE Trans. Power Apparatus & Systems,* Vol. 82, pp. 291–297, June 1963. (Sections 2.1, 2.3, 3.1, and 3.5; Figs. 2.1, 2.3, 2.12, and 3.3.)

©1963 IEEE. Reprinted with permission from C. E. Gunn, "Improved Starting Performance of Wound-Rotor Motors Using Saturistors," *IEEE Trans. Power Apparatus & Systems,* Vol. 82, pp. 298–302, June 1963. (Sections 2.1, 2.3, 3.1, and 3.5; Figs. 2.1, 2.3, 2.12, and 3.3.)

©1979 IEEE. Reprinted with permission from J. E. Leiss, "Induction Linear Accelerators and Their Applications," *IEEE Trans.* NS. Vol. 26, No. 3, pp. 3870–3876, June 1979. (Section 3.11.)

©1981 IEEE. Reprinted with permission from K. Denno, "Synchronized Switching Mechanism for the Linear Induction Accelerators," *IEEE Trans.* NS. Vol. 28, No. 3, pp. 3073–3075, June 1981. (Section 3.11.)

©1985 IEEE. Reprinted with permission from K. Denno, "Damping of Inrush Current in Synchronous Generator During State of Induction," *Proc. IEEE Society of Industrial Applications Conference,*

Toronto, Canada, pp. 854–858, 1985. (Section 3.10; Figs. 3.11, 3.12, 3.13, and 3.14.)

©1977 IEEE. Reprinted with permission from K. Denno, "Feasibility of AC Induction Generator with Hard Ferromagnetic Core in the Rotor," *Proc. IEEE International Electrical, Electronics Conf. and Exposition,* pp. 174–175, Sept. 1977. (Section 3.8.)

©1975 IEEE. Reprinted with permission from K. Denno, "Eddy-Current Theory in Hard, Thick Ferromagnetic Materials," *IEEE Conf. Paper No. C75-005-4,* pp. 1–6, IEEE-PES Winter Meeting, 1975. (Section 2.6; Figs. 2.24–2.29.)

©1987 Coil Winding Association Proc. Reprinted with permission from K. Denno, "Control of Hysteresis Motor Through Saturistor in the Rotor," *Coil Winding Association Proc.,* pp. 27–32, Rosemont, Illinois. (Section 3.7; Figs. 3.4 and 3.50.)

©1984 IASTED. Reprinted with permission from K. Denno, "Three Dimensional Model of Eddy-Current Theory in Hard Ferromagnetic Material," *Proc. IASTED International Symposium,* pp. 16–18, Nov. 1984. (Section 2.8.)

©1976 IEEE. Reprinted with permission from K. Denno, "Eddy-Current Theory for Hard Ferromagnetic Core Reactor with Exact Magnetization Curve," *IEEE. October 1976 Power Symposium Record,* Kansas State University, Manhattan, Kansas. (Section 2.7; Figs. 2.31–2.35.)

©1991 CRC Press, Boca Raton, Florida. Reprinted with permission from K. Denno, *High Voltages Engineering in Power Systems* [Eqs. (4.11), (4.14), (4.25), (4.26), (4.30), and (4.33).]

©1984 IASTED. Reprinted with permission from K. Denno, "Mathematical Modelling of Deep-Bar and High Impedance Rotor Induction Motor with Hard Magnetic Core," *Proc. IASTED International Symposium,* New Orleans, pp. 13–15, Nov. 1984. (Sections 2.7 and 3.6.)

©1986 EPRI. Reprinted with permission from "Sealed In Silicon (The Power Electronics Revolution)," published in *EPRI Journal,* pp. 5–15, Dec. 1986 by John Douglas, Science Writer. Technical information was provided by Narain Hungorani and Harshad Mehta of Electrical Systems Division, Ralph Ferraro of Energy Management and Utilization Division, and Frank Goodman of Advanced Power Systems Division. (Section 10.1.)

©1989 EPRI. Reprinted with permission from "The Future of Transmission (Switching to Silicon)," published in *EPRI Journal,* pp. 5–13, June 1989 by John Douglas, Science Writer. Technical information was provided by Narain Hungorani, Frank Young, and Robert Iveson of Electrical Systems Division. (Section 10.2.)

Electric Protective Devices

Protective Role of Resistances, Coils, and Condensers

Lumped components of resistances, coils, and condensers have been used to provide required protection at various places in power system networks involving transmission lines, junctions of lines where surge impedance changes, the exit of generating power station, at the entry or exit of transformer substations, in conjugation with or without spark arrestor to supplement protection from effects of lightning surge, in series or across a switching gear for a transmission line, or any other power system device. In various textbooks and literature materials, analytical as well as design-subjective matter have been presented with clear, practical, effective applicational benefits for all those circuit elements in place.

In this chapter, the author preferred to deal with this traditional subject—namely, the protective role of resistances, coils, and condensers in power system installations—by presenting solutions of selected principle cases involving those lumped parameter circuit elements, rather repeating theories regarding their characteristics, transmission lines rules of reflection and transmission, the principles of series and parallel resonance, effect of line terminations, and the role of cables as protective element. In the cases to be presented, the principles of continuity of electric current, voltage, and surge impedances will be applied, as well as the conventional rules of reflection and transmission.

1.1 On Transmission Line Representation

Any transmission line can be divided over a large number of T-sections as shown in Fig. 1.1.

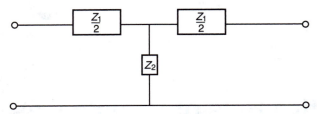

Figure 1.1 T-section of transmission line.

Z_{a-b} with c-d open is:

$$Z_{a-b \atop o-c} = Z_{oc} = \frac{Z_1}{2} + Z_2 \qquad (1.1)$$

Z_{a-b} with c-d short-circuited is:

$$Z_{a-b \atop s-c} = \frac{Z_1}{2} + \frac{Z_1 Z_2}{Z_1 + 2Z_2} \qquad (1.2)$$

If Z_0 is the surge impedance of the line, then

$$Z_0 = \sqrt{Z_{0-c} Z_{s-c}} \qquad (1.3)$$

Also, if $Z_2 = \dfrac{1}{Y_2} = \dfrac{1}{Y}$, and $Z_1 = Z_0$. Therefore,

$$Z_0 = \sqrt{\frac{Z}{Y}} \qquad (1.4)$$

and

$$\gamma = \alpha + \beta \text{ as the propagation constant}$$
$$= \sqrt{ZY} \qquad (1.5)$$

1.2   On the Status of Damping Involving *RLC* Circuit

On the status of damping involving RLC network, Fig. 1.2 shows an RLC series combination containing a charged capacitance with voltage E.

Therefore, at the moment of $t = O^+$, after closing the switch S, we can write the following:

$$Ri(t) - L\frac{di(t)}{dt} - \frac{1}{c}\int i(t)dt = 0 \qquad (1.6)$$

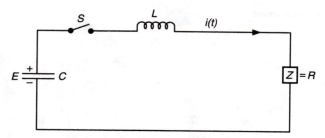

Figure 1.2 *RLC* circuit with charged capacitance.

or

$$\frac{d^2i}{dt^2} - \frac{R}{L}\frac{di}{dt} + \frac{i}{Lc} = 0 \tag{1.7}$$

Solution of Eq. (1.7) is, in general:

$$i(t) = A_1 e^{m_1 t} + A_2 e^{m_2 t} \tag{1.8}$$

where

$$m_1, m_2 = -\frac{R}{2L} = \sqrt{\left(\frac{R}{2L}\right)^{2} - \frac{1}{Lc}} \tag{1.9}$$

Underdamped case

$$\frac{1}{Lc} \ll \frac{R}{2L} \tag{1.10}$$

Therefore,

$$m_1, m_2 = -\frac{R}{2L} \pm \frac{j}{\sqrt{LC}}$$

then

$$i(t) = e^{-\frac{Rt}{LC}}\left[A_1 e^{-j\frac{t}{\sqrt{LC}}} + A_2 e^{j\frac{t}{LC}}\right]$$

$$\text{where at } t = 0^+, i = 0 \tag{1.11}$$

And, since $\int i(t)dt = E$; therefore

$$A_1 = -A_2 = -\frac{E}{LC} = A \tag{1.12}$$

Hence,

$$i(t) = \frac{E}{\sqrt{LC}}\, e^{-\frac{Rt}{\sqrt{LC}}}\, sin\, \frac{t}{LC} \tag{1.13}$$

Critical damped case occurs when

$$\frac{R}{2L} = \frac{1}{LC}$$

Therefore,

$$m_1 = M_2 = -\frac{R}{2L}$$

Therefore,

$$i(t) = Ae^{-\frac{Rt}{L}} = E\backslash Re^{-\frac{Rt}{L}} \qquad (1.14)$$

Overdamped case, occurs when

$$\left(\frac{R}{2L}\right)^2 > \frac{1}{2c}$$

Hence,

$$\sqrt{\left(\frac{R}{2L}\right)^2 - \frac{1}{LC}} = M$$

Therefore,

$$m_1, m_2 = -\frac{R}{2L} \pm M = m'_1, m'_2$$

and

$$i(t) = A_1 e^{m'_1 t} + A_2 e^{-m'_2 t} \qquad (1.15)$$

If m'_1 is positive, $A_1 = 0$, and then

$$i(t) = \frac{E}{R} e^{-m'_2 t} \qquad (1.16)$$

1.3   On Cables and Distortionless Lines

For cables, their inductance and conductance is very small, to the extent that it could be neglected.

Therefore, $Z = R$ and $Y = j\omega c$. The surge impedance is expressed by

$$\sqrt{\frac{Z}{Y}} = \sqrt{\frac{R}{j\omega c}}$$

Therefore,

$$Z_0 = \sqrt{\frac{R}{\omega C}}\left(\frac{\angle \pi}{2}\right)$$

$$\text{and } \gamma = \sqrt{R\omega C}\left(\frac{\angle \pi}{2}\right) \qquad (1.17)$$

For distortionless lines, there will be no frequency or phase delay, implying equal phase angle for both their series impedance and shunt admittance, that is,

$$\frac{\omega C}{G} = \frac{\omega L}{R} \qquad (1.18)$$

or

$$LG = RC \qquad (1.19)$$

since

$$Z = R + j\omega L$$

$$Y = G + j\omega C$$

Therefore, from Eqs. (1.18) and (1.19)

$$Y = \frac{GZ}{R} = \frac{CZ}{L} \qquad (1.20)$$

γ is the propagation constant

$$\gamma = \alpha + \beta$$

$$= (R + j\omega L)\sqrt{\frac{G}{R}} \qquad (1.21)$$

Therefore,

$$\alpha = \sqrt{\frac{C}{L}}\,R \qquad (1.22)$$

$$\beta = \omega\sqrt{LC} \qquad (1.23)$$

v as the velocity of propagation is given by

$$= \omega/\beta = \frac{1}{\sqrt{LC}} \qquad (1.24)$$

Z_0 as the surge impedance is given by

$$= \sqrt{\frac{Z}{Y}} = \sqrt{\frac{R}{G}} = \sqrt{\frac{L}{C}} \qquad (1.25)$$

1.4   On the Protective Role of a Shunt Resistance Preceding a Group of Lines

In Fig. 1.3, a transmission line of surge impedance Z_1 is connected at a junction to a set of n lines, each having a surge impedance of Z_2. Also at the junction, a protective drain resistance is shunted to ground.

Calculate the design condition on R for minimum i^2R.

solution

Let

$$Z_2' = \frac{Z_2}{n} \qquad (1.26)$$

The interconnection of Fig. 1.3, therefore, becomes as shown in Fig. 1.4.

Continuity of voltages require

$$e_1 = e_2 \qquad (1.27)$$

and

$$i = e_2/R \qquad (1.28)$$

Let e_{t1} = the forward moving voltage surge

e_{t1} = the reflected voltage surge

e_{t2} = the transmitted voltage surge

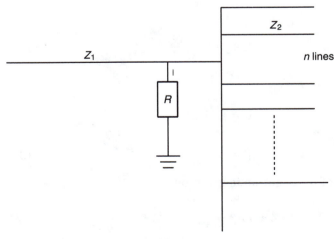

Figure 1.3 Protective role of shunt resistance.

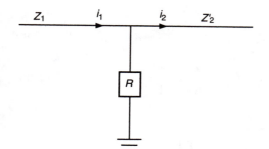

Figure 1.4 Protective shunt resistance.

Therefore, continuity of voltages at the junction requires:

$$e_{t1} + e_{r1} = e_{t2} \tag{1.29}$$

and continuity of current at the junction requires:

$$i_1 = i_2 + \frac{e_{t2}}{R} \tag{1.30}$$

or

$$\frac{e_{t1}}{Z_1} - \frac{e_{r1}}{Z_1} = \frac{e_{t2}}{Z_2'} + \frac{E_{t2}}{R} \tag{1.31}$$

Multiplying Eq. (1.31) by Z_1 and adding to Eq. (1.29) and rearranging, we obtain the following:

$$e_{t2} = \frac{\dfrac{2\,e_{t1}}{Z_1}}{\dfrac{1}{Z_1} + \dfrac{1}{Z_2'} + \dfrac{1}{R}} \tag{1.32}$$

Power dissipated in R:

$$W_R = \frac{e_{t2}^2}{R}$$

$$= \frac{\dfrac{4\,e_{t1}^{\,2}}{Z_1^{\,2}\,R}}{\left[\dfrac{1}{Z_1} + \dfrac{1}{Z_2'} + \dfrac{1}{R}\right]^2} \tag{1.33}$$

$$e_{t1} - e_{r1} = e_{t2}\,\frac{Z_1}{Z_2'} + \frac{Z_1}{R}\,e_{t2} \tag{1.34}$$

From continuity of voltages,

$$e_{t1} + e_{r1} = E_{t2} \tag{1.35}$$

$$2e_{t1} = e_{t2}\left[\frac{Z_1}{Z_2'} + \frac{Z_1}{R} + 1\right] \tag{1.36}$$

$$e_{t2} = \frac{2e_{t1}}{\left(\dfrac{Z_1}{Z_2'} + \dfrac{Z_1}{R} + 1\right)}$$

$$= \frac{\dfrac{2}{Z_1}e_{t1}}{\dfrac{1}{Z_1} + \dfrac{1}{Z_2'} + \dfrac{1}{R}} \tag{1.37}$$

The incident power W is expressed by

$$W = \frac{E^2_{t1}}{Z_1} \tag{1.38}$$

And therefore,

$$\frac{W_R}{W} = \frac{\dfrac{4\,e_{t1}{}^2}{Z_1{}^2\,R}}{\left[\dfrac{1}{Z_1} + \dfrac{1}{Z_2'} + \dfrac{1}{R}\right]^2}\frac{Z_1}{e_{t1}{}^2}$$

$$= \frac{\dfrac{4}{Z_1 R}}{\left[\dfrac{1}{Z_1} + \dfrac{1}{Z_2'} + \dfrac{1}{R}\right]^2} \tag{1.39}$$

Upon differentiating W_R/W with respect to R and equating the result to zero, the following value of R for minimum ratio of W_R/W is given:

$$R = \frac{Z_1 Z_2'}{Z_1 + Z_2'} \tag{1.40}$$

Design condition on R by Eq. (1.40) is the shunt combination of the incoming line surge impedance with the outgoing surge impedance.

1.5 On the Design Condition of a Series Resistance Inserted between an Incoming Line and a Group of Lines

See Fig. 1.5. The equivalent surge impedance Z_2' of n parallel lines, each of surge impedance Z_2, is given by

$$Z_2' = \frac{Z_2}{n} \tag{1.41}$$

The equivalent circuit may now be drawn as shown in Fig. 1.6.

The protective resistance R is assumed to be lumped, i.e., concentrated in a space short in comparison with that over which the pulse extends.

The resistance R, series connected, leaves the current summation at the junction unaltered, that is,

$$i_1 = i_2 \tag{1.42}$$

However, the voltage in the two lines differs by the voltage drop in the resistance. Therefore,

$$e_1 = e_2 + Ri_2 \tag{1.43}$$

We now decompose e_1 onto forward traveling and rearward traveling voltage pulses and consider that lines on the right-hand side only contain transmitted voltage pulse.

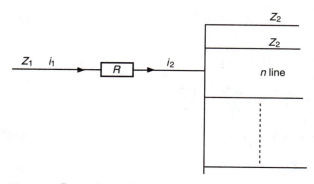

Figure 1.5 Protective series resistance.

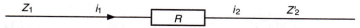

Figure 1.6 Protective series resistance.

Therefore,

$$e_{t1} + e_{r1} = i_2 R + e_{t2} \tag{1.44}$$

Now, from Eq. (1.43),

$$\frac{e_{t1} - e_{r1}}{Z_1} = \frac{e_{t2}}{Z_2'} \tag{1.45}$$

Multiplying Eq. (1.45) by Z_1 and adding Eq. (1.44), we get (by rearranging):

$$e_{t2} = \frac{2Z_2'}{Z_1 + Z_2' + Re_{t1}} \tag{1.46}$$

From Eq. (1.45) we get e_{r1}, by substituting for e_{t2} as obtained in Eq. (1.46), where

$$e_{r1} = \frac{Z_2' - Z_1 - R}{Z_2' + Z_1 + R} e_{t1} \tag{1.47}$$

Power loss in the resistance R is given by

$$W_R = Ri_2^2 = R\,\frac{e_{t2}^2}{Z_2'} \tag{1.48}$$

Power contained in the incident pulse is given by

$$W = \frac{e_{t1}^2}{Z_1} \tag{1.49}$$

From Eqs. (1.48) and (1.49) we get

$$\frac{W_R}{W} = \left(\frac{Re_{t2}^2}{Z_2'}\right) \div \left(\frac{e_{t1}}{Z_1}\right)^2 \tag{1.50}$$

Substituting for e_{t2} from Eq. (1.44)

$$\eta = \frac{W_R}{W}$$

$$= \frac{4R(Z_2'^2)e_{t1}^2}{Z_2'^2(Z_1 - Z_2' + R)^2} \times \frac{Z_1^2}{Z_1 e_{t1}^2}$$

$$= \frac{4RZ_1}{(Z_1 + Z_2' - R)^2} \tag{1.51}$$

For the efficiency η expressed by W_R/W should be minimum and for the design value of R, we now write:

$$\frac{d\eta}{dR} = 4Z_1 \frac{d}{dR}\left[\frac{R}{(Z_1 + Z_2' + R)^2}\right]$$

Let $\qquad\qquad u = R \therefore \frac{du}{dR} = 1$

$$v = (Z_1 + Z_2' + R)^2 \therefore \frac{dv}{dR} = 2(Z_1 + Z_2' + R)$$

Therefore,

$$\frac{d}{dR}\left[\frac{R}{(Z_1 + Z_2' + R)^2}\right] = \frac{2(Z_1 + Z_2' + R) \times 1 - R \times 2(Z_1 + Z_2' + R)}{(Z_1 + Z_2' + R)^4}$$

$$= \frac{2(Z_1 - Z_2' + R)(1 - R)}{(Z_1 + Z_2' + R)^4}$$

$$= 0$$

$$\Rightarrow R = |(Z_1 + Z_2')| \text{ for } \eta_{min} \qquad\qquad (1.52)$$

Substituting for Z_2' from Eq. (1.43), we get

$$R = Z_1 + \frac{Z_2}{n}$$

Minimum value of efficiency η could be expressed by substituting. $R = Z_1 - Z_2$ in Eq. (1.51). Therefore,

$$\eta_{min} = \frac{4RZ_1}{(Z_1 + Z_2' + Z_1 + Z_2')^2}$$

$$= \frac{4Z_1(Z_1 + Z_2')}{4(Z_1 + Z_2)^2}$$

$$= \frac{Z_1}{(Z_1 + Z_2')}$$

or

$$\eta_{min} = \frac{Z_1}{\left(Z_1 + \dfrac{Z_2}{n}\right)}$$

$$= \frac{nZ_1}{Z_1 + Z_2} \qquad\qquad (1.53)$$

1.6   On the Protective Role of a Shunt Resistance in Series with a Linear Arrester

In Fig. 1.7, the set of lines to the right of the protective device represent surge impedances that are increasing according to an arithmetical progression.

The system to the right of point o is n lines with gradual increase in individual surge impedance with a total parallel equivalent of Z_2.

$$Z_2 = \cfrac{1}{\dfrac{1}{Z} + \dfrac{1}{Z + \Delta Z} + \dfrac{1}{Z + 2\Delta Z} + \cdots + \dfrac{1}{Z + n\Delta Z}}$$

$$= \frac{1}{Y_2}$$

$$Y_2 = \frac{1}{Z} + \frac{1}{Z + \Delta Z} + \cdots \frac{1}{Z + n\Delta Z}$$

$$= \frac{1}{Z} + \left[\cfrac{1}{1 - \dfrac{\Delta Z}{Z}} + \cdots + \cfrac{1}{1 + n\dfrac{\Delta Z}{Z}} \right] \tag{1.54}$$

e_0 is the arrester limiting voltage.

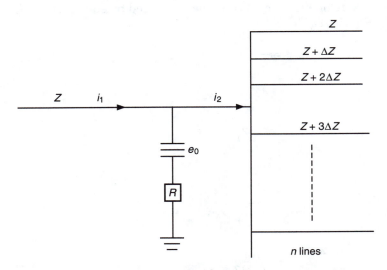

Figure 1.7 Protective linear arrester with resistance.

Practically, ΔZ is small if all the individual Zs are those for homogeneous but different transmission lines and cables. But if they are due to cables, lines, transformers, etc., ΔZ could be even greater than Z.

However, as a specific case, consider all Z's are those of homogeneous but different lines that $\delta Z \ll Z$

Therefore,

$$Y_2 \approx \frac{1}{Z}\left[1 + \left(1 - \frac{\Delta Z}{Z}\right) + \left(1 - \frac{2\Delta Z}{Z}\right) + \right.$$

$$\left.\left(1 - \frac{3\Delta Z}{Z}\right) - \cdots + \left(1 - \frac{(n-1)\Delta Z}{Z}\right)\right]$$

$$Z_2 = \frac{1}{Y_2} \tag{1.55}$$

Then, according to the result obtained:

$$\frac{1}{R_{opt}} = \frac{1}{Z'} + \frac{1}{Z_2} \tag{1.56}$$

$$Z_2 \cong \frac{1}{Z}\left[n - \frac{\Delta Z}{Z}\left(\frac{(n-1)+1}{2}\right)\right]$$

$$\approx \frac{n}{Z}\left[1 - \frac{1}{2}\frac{\Delta Z}{Z}\right]$$

$$\approx \frac{n}{Z}\left[\frac{2Z - \Delta Z}{2Z}\right]$$

$$R_{opt} \cong 1\left[\frac{1}{Z'} - \frac{2Z^2}{n(2Z - \Delta Z)}\right] \tag{1.57}$$

Equation (1.57) is valid for a homogeneous system of small increase in Z.

1.7 On the Protective Role of Condenser

Figure 1.8 shows a condenser of capacitance C inserted between two transmission lines of surge impedances Z_1 and Z_2.

Ohmic value from Z_1 and Z_2 associating itself for the time constant of the network is their parallel combination, namely,

$$Z_1 \| Z_2 = \frac{Z_1 Z_2}{Z_1 + Z_2} = Z_{eq}$$

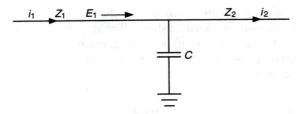

Figure 1.8 Protective shunt capacitance.

Therefore,

$$\tau_c = CZ_{eq}$$

The distance an incident surge will travel in the duration of τ_c is equal to $\tau_c v_0$, where v_0 is the velocity of propagation. Current surge transmitting through Z_2 is:

$$i_{t2} = \frac{2Z_1 I_0}{Z_1 - Z_2}\left[1 - e^{-\frac{t}{\tau_c}}\right] \tag{1.58}$$

where

$$I_0 = \frac{E_1}{Z_1}$$

The reflected current is as follows:

$$i_{r1} = \frac{e_{r1}}{Z_1}$$

$$= -E + \frac{2Z_2 E}{Z_1 + Z_2}\left[1 - e^{-\frac{t}{\tau_c}}\right] \tag{1.59}$$

1.8   On the Solution of Transmitted Voltage Surge Due to an Incident-Strong Sinusoidal Surge Confronted with Protective *L* or *C*

In Fig. 1.9 a powerful voltage surge e_{t1} is incident on a transmission line whose surge impedance is Z_1, followed by either a protective lumped element of either an inductance L or capacitance C.

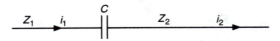

Figure 1.9 Protective series capacitance.

The intention here is to obtain a solution for the transmitted voltage surge e_{t2} in line 2 through the classical approach of mathematical tools.

$$e_{t1} = E\cos(\omega t + \psi) \tag{1.60}$$

From the principles of continuity,

$$i_1 = i_2 \tag{1.61}$$

and

$$e_1 = e_2 + L\,\frac{di_2}{dt} \tag{1.62}$$

since

$$i_2 = \frac{e_{t2}}{Z_2} \tag{1.63}$$

Therefore, we can write Eq. (1.62) as follows:

$$e_{t1} - e_{r1} = e_{t2} + \frac{L}{Z_2}\,\frac{de_{t2}}{dt} \tag{1.64}$$

From Eq. (1.61):

$$i_1 = \frac{e_{t1} - e_{r1}}{Z_1} \quad \text{and} \quad i_2 = \frac{e_{t2}}{Z_2}$$

Therefore,

$$\frac{e_{t1} - e_{r1}}{Z_1} = \frac{e_{t2}}{Z_2}$$

or

$$e_{t1} - e_{r1} = \frac{Z_1}{Z_2}\,e_{t2} \tag{1.65}$$

Adding Eq. (1.64) and $Z_1 \times$ Eq. (1.65), we obtain:

$$\frac{de_{t2}}{dt} - \frac{Z_1 + Z_2}{L}\,e_{t2} = \frac{2Z_2}{L}\,e_{t1} \tag{1.66}$$

Introducing the time constant

$$\tau = \frac{L}{Z_1 + Z_2}$$

into Eq. (1.66), the result is:

$$\mathrm{T}\,\frac{de_{t2}}{dt} + e_{t2} = \frac{2Z_2}{Z_1 + Z_2}\,e_{t1} \tag{1.67}$$

Substituting e_{v1} from Eq. (1.60) into Eq. (1.67),

$$\frac{de_{t2}}{dt} + \frac{1}{\tau} e_{t2} = \frac{2Z_2E}{\tau(Z_1 + Z_2)} \, cos(\omega t + \psi) \tag{1.68}$$

Let

$$a = \frac{1}{\tau}$$

$$b = \frac{2Z_2E}{\tau(Z_1 + Z_2)}$$

Therefore, Eq. (1.68) becomes:

$$\frac{de_{t2}}{dt} - ae_{t2} = bcos(\omega t + \psi) \tag{1.69}$$

With $D = \frac{\delta}{\delta\tau}$, Eq. (1.69) becomes:

$$(D + a)e_{t2} = bcos(\omega t + \psi)$$

The characteristic equation is:

$$D + a = 0$$
$$D = -a$$

The complementary solution is:

$$e_{t2} = c_1 e^{-at}$$

The particular integral of solution is:

$$e_{t2-p} = Acos(\omega t + \psi) + Bsin(\omega t + \psi) \tag{1.70}$$

Substituting Eq. (1.70) into Eq. (1.69) and comparing coefficients, we obtain:

$$A = \frac{ab}{a^2 + w^2}$$

$$B = \frac{wb}{a^2 + w^2} \tag{1.71}$$

Total solution for $e_{t2} = e_{v2-c} + e_{v2-p}$. Therefore,

$$e_{t2} = c_1 e^{-at} + \frac{ab}{a^2 + \omega^2} \, cos(\omega t + \psi) + \frac{\omega b}{a^2 + \omega^2} \, sin(\omega t + \psi) \tag{1.72}$$

At $t = 0$, $e_{t2} = 0$. Therefore,

$$0 = c_1 - \frac{ab}{a^2 + \omega^2} \cos\psi + \frac{\omega b}{a^2 + \omega^2} \sin\psi$$

The general solution for e_{t2} is:

$$e_{t2} = \frac{b}{a^2 - \omega^2}[a \cos(\omega t + \psi) + \omega \sin(\omega t + \psi) - (a\cos\psi + \omega\sin\psi)e^{-at}]$$

$$= \frac{b}{a^2 - \omega^2} [\sqrt{a^2 + \omega^2} \cos(\omega t + \psi - \theta) - \sqrt{a^2 + \omega^2}\cos(\psi - \theta)] \qquad (1.73)$$

where

$$\theta = tan^{-1} \frac{\omega}{a} \qquad (1.74)$$

or

$$e_{t2} = \frac{b}{\sqrt{a^2 + \omega^2}} [\cos(\omega t + \psi) - \cos\omega e^{-at}] \qquad (1.75)$$

$$\Phi = \psi - \theta \qquad (1.76)$$

Substituting a and b into Eq. (1.75):

$$e_{t2} = \frac{2EZ_2}{\tau(Z_1 + Z_2)\omega^2 + \sqrt{\dfrac{1}{\tau^2}}} \left[\cos(\omega t + \psi) - \cos\psi e^{-\frac{t}{\tau}}\right] \qquad (1.77)$$

$$e_{t2} = \frac{2EZ_2}{\tau(Z_1 + Z_2)\sqrt{\dfrac{1 + \omega^2\tau^2}{\tau^2}}} \left[\cos(\omega t + \psi) - \cos\psi e^{-\frac{t}{\tau}}\right]$$

$$= \frac{2Z_2}{Z_1 + Z_2} \frac{E}{\sqrt{1 + \omega^2\tau^2}} \left[\cos(\omega t + \psi) - \cos\psi e^{-\frac{t}{\tau}}\right] \qquad (1.78)$$

1.9   On the Continuity of Surge Impedances at a Junction

In reference to Fig. 1.10, at the junction, continuity for currents and voltages implies:

$$i_1 = i_2$$

$$e_1 = e_2$$

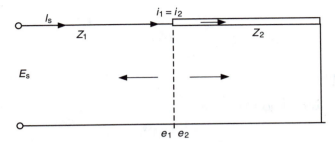

Figure 1.10 Continuity of surge impedances.

$$e_1 = i_1 Z_{in1}$$

$$e_2 = i_2 Z_{in2}$$

Therefore,

$$Z_{in1} = Z_{in2}$$

From general transmission line theory, the input impedance for any line termination is given by:

$$Z_{in} = \frac{E_s}{I_s} = \frac{Z_0[Z_r cosh\gamma l + Z_0 sinh\gamma l]}{[Z_0 cosh\gamma l + Z_r sinh\gamma l]} \tag{1.79}$$

$Z_1 n_1$ for open-circuit termination can be found by using $Z_0 = \chi$. Therefore,

$$Z_{in1} = \frac{Z_0 cosh\gamma l + \dfrac{Z_0}{Z_r} sinh\gamma l}{sinh\gamma l + \dfrac{Z_0}{Z_r} cosh\gamma l}$$

$$= Z_0 \frac{cosh\gamma l}{sinh\gamma l}$$

$$= Z_1 ctnh\gamma l \tag{1.80}$$

And $Z_{in2} = Z = 0$ represents the case of short-circuit termination.

$$Z_{in2} = Z_2 \frac{sinh\gamma l}{cosh\gamma l} = Z_2 tanh\gamma l \tag{1.81}$$

Therefore,

$$Z_1 ctnh\gamma l = Z_2 tanh\gamma l$$

and for

$$\gamma = j\beta$$

$$Z_1 ctnh \ j \ \beta l = Z_2 tanh \ j \ \beta l$$

and

$$Z_1 ctn \ \beta_1 a_1 = Z_1 tan \ \beta_2 a_2 \qquad (1.82)$$

where

$$v = \frac{v}{\beta}$$

or

$$\beta = \frac{v}{v}$$

Therefore,

$$Z_1 ctn \ \frac{va_1}{v_2} = Z_2 tan \ \frac{va_2}{v_2} \qquad (1.83)$$

If the velocity of propagation is continuous:

$$v_1 = v_2$$

and

$$tan^2 \ \frac{va}{v} = \frac{Z_1}{Z_2}$$

or

$$cos \ \frac{va}{v} = \pm \ \cfrac{1}{1 + \sqrt{\cfrac{Z_1}{Z_2}}} \qquad (1.84)$$

with

$$Z_2 = \frac{Z_1}{2}$$

$$cos \ \frac{va}{v} = \pm \ \frac{1}{\sqrt{3}} \qquad (1.85)$$

$$\frac{va}{v} = cos^{-1} \pm \ \frac{1}{\sqrt{3}}$$

$$= 0.956. \ 2.186. \ 4.098. \ 5.328 \qquad (1.86)$$

1.10   On the Continuity of Open-Circuit and Short-Circuit Impedances and Identification of Natural Frequencies

At junction (0) of Fig. 1.11, equating Z_{in} of the open-circuit line with surge impedance $Z\backslash 2$, and that of the short-circuit line with surge impedance Z, therefore,

$$Z_{in}\,(o.c) = \frac{Z}{2}\,ctnh\gamma a \qquad (1.87)$$

and

$$Z_{in}(s.c) = Ztanh\gamma a \qquad (1.88)$$

Since no losses are considered, therefore, $\gamma = j\beta$ and the hyperbolic elements will be replaced with tri-geometrics:
Therefore,

$$\frac{1}{2}\,Zctn\beta a = Ztan\beta a \qquad (1.89)$$

or

$$tan^2\beta a = \frac{1}{2}$$

$$\beta = \frac{v}{\mathrm{v}}$$

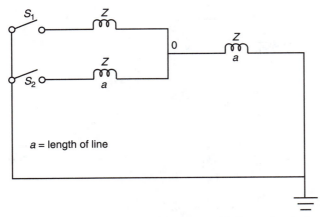

a = length of line

Figure 1.11 Continuity of open circuit and short circuit impedances.

Where v is the wavelength. and v is the velocity of propagation. Therefore,

$$tan^2 \frac{va}{v} = \frac{1}{2}$$

$$cos \frac{va}{v} = \pm \frac{1}{\sqrt{1 + \dfrac{1}{2}}}$$

$$= \pm \sqrt{\frac{2}{3}} \qquad (1.90)$$

$$\frac{va}{v} = cos^{-1}\left(\pm \sqrt{\frac{2}{3}}\right)$$

$$= 0.618.\ 2.53.\ 3.76.\ \ldots$$

and

$$f = \frac{v}{2\pi}$$

$$= \frac{0.618v}{2\pi a},\ \frac{2.53v}{2\pi a},\ \frac{3.76v}{2\pi a} \cdots \qquad (1.91)$$

1.11   On the Sudden Switching Off of a Transmission Line Having Lumped Self-Inductance and Capacitance

Initial conditions:

Current in the line (at $t = O^+$) = I amp

Time interval for switching = T μ sec

Look into calculations of the following:

■ Front-rise length of the interrupted voltage surge:

$$e = v_0\tau\ km \qquad (1.92)$$

where v_0 is the velocity of propagation.

- Magnitude of switching surge:

$$V_s = 2ZI \tag{1.93}$$

where Z is the surge impedance of the line. Usually, for overhead lines $Z \rightarrow 500 \ \Omega$, for cables $Z \rightarrow 50 \ \Omega$, and for transformer $Z \rightarrow 5,000 \ \Omega$.

For ordinary steady-state system operating at 60 Hz, the wave spread $\lambda - 5,000$ km. This implies that increase in the switching time T will result in extensive spread at the switching surge over a long length of the line and, consequently, weaken it.

1.12   On the Impedance Transformation Theorem

This theorem states that if a set of four terminal networks containing only pure reactances is interconnected between a generator with its internal impedance and any load, then, if at any section the impedance in one direction is the conjugate of that impedance in opposite direction, there will be a conjugate match of impedances at every other section in the system.

At any of sections A-A′, B-B′, C-C′, and D-D′ (shown in Fig. 1.12), conjugate matching exists such that $Z_{A\text{-}A'\text{-}left} = Z_{A\text{-}A'\text{-}right}$, and the same could be said at sections B-B′, C-C′, and D-D′.

Now, in Fig. 1.13, Z_1, Z_2, and Z_3 are unsymmetrical; Z_{l1} and Z_{l2} could be called image impedances if $Z_{1\text{-}2}$ at one direction is equal to $Z_{3\text{-}4}$ at the other direction. Therefore,

$$Z_{l1} = Z_1 + \frac{(Z_2 + Z_{l2})Z_3}{Z_2 + Z_3 + Z_{l2}} \tag{1.94}$$

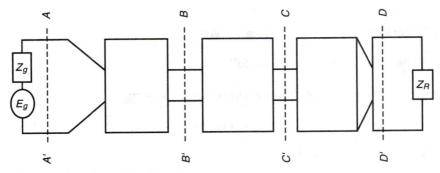

Figure 1.12 Impedance transformation.

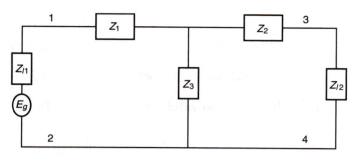

Figure 1.13 Image impedances.

$$Z_{l2} = Z_2 + \frac{(Z_1 + Z_{l1})Z_3}{Z_1 + Z_3 + Z_{l1}} \tag{1.95}$$

Clearing fractions from Eqs. (1.94) and (1.95), we have:

$$Z_{l1}(Z_2 - Z_3) + Z_{l1}Z_{l2} = Z_1Z_2 + Z_2Z_3 + Z_1Z_3 + Z_{l2}(Z_1 + Z_3) \tag{1.96}$$

and

$$Z_{l2}(Z_1 + Z_3) + Z_{l1}Z_{l2} = Z_1Z_2 + Z_2Z_3 + Z_1Z_3 + Z_{l1}(Z_2 + Z_3) \tag{1.97}$$

Solving for Z_{l1} and Z_{l2} from equations (1.96) and (1.97):

$$Z_{l1} = \sqrt{\left(\frac{Z_1 + Z_3}{Z_2 + Z_3}\right)(Z_1Z_2 + Z_2Z_3 + Z_1Z_3)} \tag{1.98}$$

$$Z_{l2} = \sqrt{\left(\frac{Z_2 + Z_3}{Z_1 + Z_3}\right)(Z_1Z_2 + Z_2Z_3 + Z_1Z_3)} \tag{1.99}$$

The image impedances can be determined in terms of the open and short-circuited impedances.

Let: Z_{01} = impedance at terminals 1-2 with 3-4 open
  Z_{02} = impedance at terminals 3-4 with 1-2 open
  Z_{51} = impedance at terminals 1-2 with 3-4 short-circuited
  Z_{52} = impedance at terminals 3-4 with 1-2 short-circuited

Therefore,

$$Z_{01} = Z_1 + Z_3 \tag{1.100}$$

$$Z_{s1} = Z_1 + \frac{Z_2Z_3}{Z_2 + Z_3} \tag{1.101}$$

$$Z_{02} = Z_2 + Z_3$$

$$Z_{s2} = Z_2 - \frac{Z_1Z_3}{Z_1 + Z_3}$$

Therefore,

$$Z_{l1} = \sqrt{Z_{01}Z_{s1}} \qquad (1.102)$$

$$Z_{l2} = \sqrt{Z_{02}Z_{s2}} \qquad (1.103)$$

Equations (1.102) and (1.103) are also valid for an equivalent Π networks.

1.13   Problems

1.1   On the protective role of a shunt resistance presented in Sec. 1.4, compare the design of equivalent R: (1) when all the n lines are overhead, and (2) when they are all cables.

1.2   On the protective role of series resistance presented in Sec. 1.5, compare the design of equivalent R: (1) when all the n lines are cables connected in parallel with a transformer whose characteristic impedance is 100 times that of a single cable, and (2) when all the n lines are overhead with the additional transformer having a characteristic impedance 10 times that of a single line.

1.3   On the protective role of a resistance in series with a lightning arrester presented in Sec. 1.6, repeat the process of obtaining expression for optimum R when the n lines to the right increase from one to the other according to a geometrical progression.

1.4   On the protective role of a shunt condenser presented in Sec. 1.7, repreat the process of obtaining a solution for the reflected current pulse for a series capacitance inserted between a cable to its left and an ungrounded transformer whose surge impedance is 100 times that of the cable.

1.5   Using the principle of continuity of Z_{input} at node O shown in Fig. 1.14 obtain a solution for the set of natural frequencies.

1.6   In Fig. 1.15, calculate the ratio of excess voltage surge with respect to an incident voltage surge of 50 kv. Also calculate possible spectrum of natural frequencies.

1.7   In Fig. 1.16, given

$$n = 8, Z/line = 500 \ \Omega$$

$$e_{t1} = 80 \ \text{kv}, e_0 = 32 \ \text{kv}$$

$$Q = 150\Omega$$

Calculate current flow in the arrester.

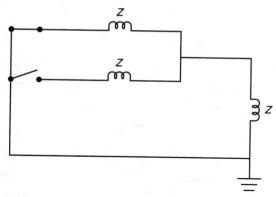

Figure 1.14 Problem 5.

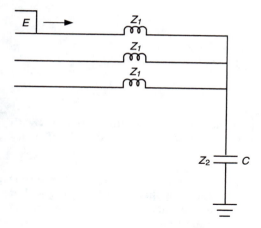

Figure 1.15 Problem 6.

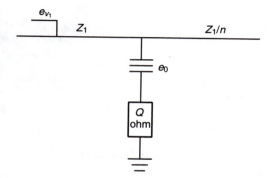

Figure 1.16 Problem 7.

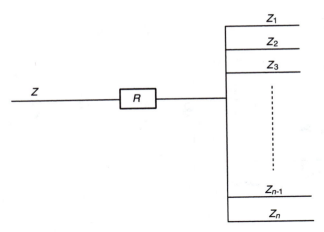

Figure 1.17 Problem 8.

1.8 In the Fig. 1.17

$$Z_1 = Z_2 = Z_3 = Z$$
$$Z_2 = \frac{Z}{n}$$
$$Z_1 = Z$$

Calculate the limiting efficiency η for the ratio of power dissipated in R with respect to total incident power, in terms of n only. Then find such efficiency when $n \to \infty$

1.9 A dissipationless transmission line is suddenly subjected to the surge-time function shown in the Fig. 1.18. This surge is applied at the sending end at $t = O^+$. Calculate the current and voltage at any time t when the line-receiving end is open-circuited. Assume zero initial conditions.

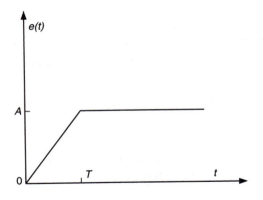

Figure 1.18 Problem 9.

1.10   A transmission line of surge impedance Z is to be connected to the following sections in separate individual cases. Comment in each case with respect to the degree of reflection, velocity of pulse propagation, and voltage transformation ratio.

1. Another line whose Z is decreasing linearly
2. Transformer system
3. Transmission line whose length is of the order of 400 km

1.11   In the foregoing system, at zero initial conditions, a step pulse is suddenly incident on the line with surge impedance Z_1. (See Fig. 1.19.) Find solutions for e_{t2}, e_{t1}, e_{r1}, and i_{r2}.

$$Z_2 \ll Z_1$$

1.12   In the foregoing transmission system (see Fig. 1.20), find:

1. The total voltage transformation ratio
2. Velocity of propagation in each region
3. Total time of pulse travel

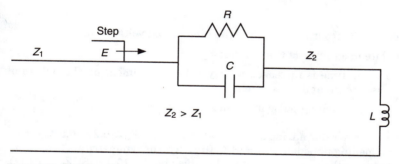

Figure 1.19   Problem 11.

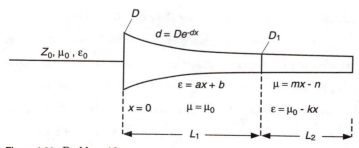

Figure 1.20   Problem 12.

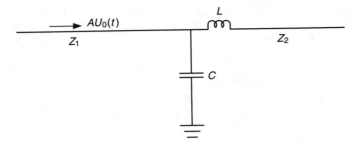

Figure 1.21 Problem 13.

1.13 An impulse $AU_0(t)$ is incident on the line with surge impedance Z_1. (See Fig. 1.21.) Find solutions for:

$$e_{t2}, i_2, e_{r1}, \text{ and } i_{r2}$$

Identify the three cases of damping in detail, with zero initial conditions.

1.14 Given a 100-kv line having completely lumped total inductance and capacitance carrying 200 A, where the line suddenly interruped in a time interval of 10^{-4} sec:

1. Calculate the front-rise length of the interrupted voltage pulse.
2. Calculate magnitude of the switching pulse.
3. If switching time is increased to, say, 0.1 sec, comment on the nature of front-rise length and pulse strength.
4. Sketch the switching pulse in 1 and 2.

1.15 An AC generator having an internal impedance $Z_g = (0.2 + j0.5)\Omega$ is connected to a loading impedance $Z_R = (8 + j10)\Omega$. The unsymmetrical (T) network existing between the generator and load is given by $Z_1 = (0.4 + j6)$, $Z_2 = (1 + j2)$, and $Z_3 = (2 + j5)$. Calculate the image impedances.

1.14 References

1. Bergen. A. R., *Power Systems Analysis,* Prentice Hall, Inc., Englewood Cliffs, N.J., 1986.
2. Denno. K., *High Voltage Engineering in Power Systems,* CRC Press, Boca Raton, Florida, 1992.
3. Everitt, W. L., *Communication Engineering,* McGraw-Hill Book Co., New York, London, 1932, 1937.
4. Harrington, R. F., *Time-Harmonic Electromagnetic Fields,* McGraw-Hill Book Co., New York, 1961.
5. Rüdenberg, Reinbold, *Electrical Shock Waves in Power Systems,* Harvard University Press, Cambridge, Massachusetts, 1968.
6. Stevenson, W. D., *Elements of Power System Analysis,* 4th ed., McGraw-Hill Book Co., New York, 1982.

Saturistor as Electromagnetic Protective Device

2.1 Generalized Properties

Currently used saturable-reactors in power systems are recognized by their ferromagnetic core associated with a normal magnetization curve passing into the origin (also known as the *B-H* curve). Replacing the ferromagnetic core by an alloy made of aluminum, cobalt, nickel, and manganese of various percentages will lead to a quite different behavior of a magnetic device known as the *saturable-resistor,* or, in short, the *saturistor.* Control for the saturation capacity of a single-phase saturistor could be accomplished by having a central ferromagnetic core with its surrounding coil biased by DC current. Figure 2.1*a, b* illustrates a simple configuration for a single-phase saturistor with DC bias control superimposed on the central leg. Figure 2.2 shows typical displaced *B-*

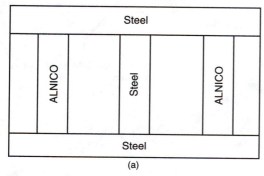

(a)

Figure 2.1 Single-phase saturister with DC biasing.

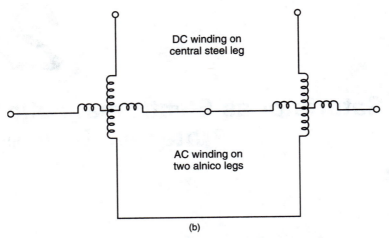

DC winding on
central steel leg

AC winding on
two alnico legs

(b)

Figure 2.1 Single-phase saturister with DC biasing. (*Continued*)

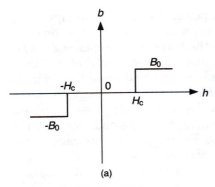

(a)

Figure 2.2 (*a*) Displaced *b-h* curve

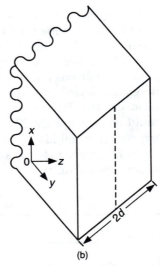

(b)

Figure 2.2 (*b*) field incidence at
saturistor.

H curve for a saturistor. The saturable with its steady-state and time-varying properties offers very tempting functions for control of current and voltage surges in power systems and electric machines.

The DC bias could be made available from the AC distribution system through rectification process.

The saturable-resistor possesses the following general properties:

1. The saturistor is a nonlinear resistor whose value can rise sharply for AC current values beyond the pickup value corresponding to the coercive force. Most of its resistive component is due to hysteresis loss with the hard magnetic core, the ALNICO, which is hysteresis material. All ohmic values X, R, and Z start to rise sharply as soon as applied mmf exceeds the coercive force. This is a unique property which is the inverse of magnetic saturation.

   The ohmic values are also frequency dependent and, in general, it can be stated that:

$$Z, X, R = \text{function of } (i_{ac}, \text{frequency})$$

2. The power factor of the saturable-resistor is in the range from $0.7 \rightarrow 0.8$ over a considerable range of AC current. Also the power factor is almost independent of frequency since hysteresis loss is proportional to frequency.

3. Level of AC saturation can be changed by DC bias superimposed on a center leg with iron core. The iron cross section is double that of each of the AC leg of ALNICO core. Therefore Z, X, R = function of $(i_{ac}, \text{frequency}, i_{dc})$.

4. To preclude transformer action in the center DC leg, the flux in it must not vary throughout the AC cycle. The AC coils may be connected in parallel so that each will carry an equal amount of magnetic flux with zero flux in the center leg. Pattern of DC flux is that it distributes itself equally over the two AC legs, injecting the same changes in each leg accordingly. However, with parallel connection of the two AC coils, any change in the control DC current will induce opposing mmfs in each of the two AC coils.

5. The two AC coils may be connected in series, but with any dissymmetry, may lead to a high induced AC voltage in the control DC leg, leading to an imposed design of few turns and thicker insulation for the center leg, and a reduced degree of control.

Steady state and transient performance of the saturable-resistor are illustrated in the following diagrams:

1. Figure 2.3 shows a (B-H) hysteresis curve for a reactor with ALNICO, showing the delayed zero flux point until the applied mmf reaches the order of about 75 percent of the peak value.

2. Figures 2.4 and 2.5 show the behavior of R, X, Z and power loss/in^3 for Star and Delta connected reactor indicating the pickup point at which magnetization starts.

3. Figures 2.6, 2.7, 2.8, and 2.9 show the surge current response and the rate of decay at various levels of DC premagnetization, and different suddenly applied AC voltages.

4. Figure 2.10 shows the behavior of the reactor time-constant as the rms value of the AC current charges.

2.2   Integral Component of the Surge Damper

In this application, the saturable-resistor could replace the nonlinear resistive element in the conventional surge damper.

The interconnection of the saturable-resistor with the SCR to act as a surge damper is shown in Fig. 2.11.

In distribution systems with normal operating voltage of the order of 13.2 KV, several SCRs could be connected in series, since SCR voltage rating is around 2.5 KV.

The idea of a surge-damper incorporating SCRs with the saturable-resistor is based on:

- Its ability for changing ohmic values automatically once the initial current surge exceeds the saturistor pickup value.

- The ohmic values of the saturistor, especially its resistance increase with the frequency. Hence, the total increase in the saturistor ohmic values will amplify as the magnitude of the surge current and frequency increase.

- It can dissipate a considerable amount of surge energy of the order of 300 to 500 watts/in^3.

- Losses inside the electromagnetic circuit could be characterized as equivalent to a form of copper loss.

The saturable-resistor with ALNICO as its core can dissipate a considerable amount of power, since such a core is a hysteresis material, in addition to eddy-current losses. Tests have shown that power dissipation varies from 300 to 500 watts per cubic inch of ALNICO. A cross section of a single-phase saturistor is shown in Fig. 2.12.

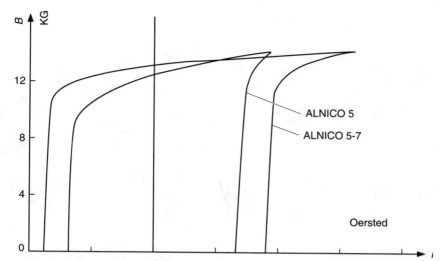

Figure 2.3 *B-H* curve of ALNICO 5-7.

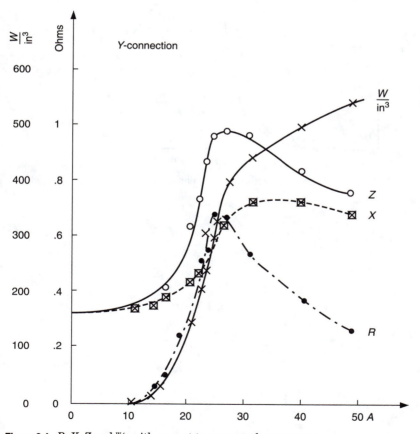

Figure 2.4 *R, X, Z* and $^W/_{in^3}$ with respect to average phase current.

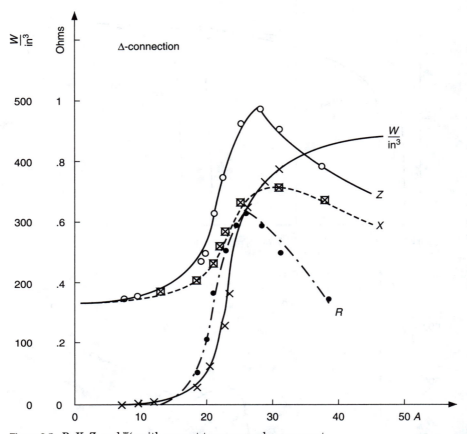

Figure 2.5 R, X, Z, and $^W/_{in^3}$ with respect to average phase current.

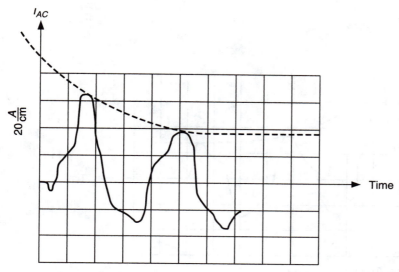

Figure 2.6 Transient response at:

$$I_{DC} = 27 \text{ amps}$$
$$I_{AC\backslash SS} = 36 \text{ amps}$$
$$V_{AC} = 57.5 \text{ volts}$$

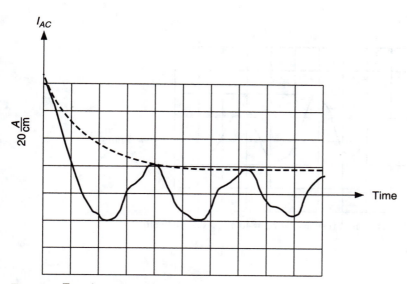

Figure 2.7 Transient response at:

$$I_{DC} = 27 \text{ amps}$$
$$I_{AC\backslash SS} = 30 \text{ amps}$$
$$V_{AC} = 51.5 \text{ volts}$$

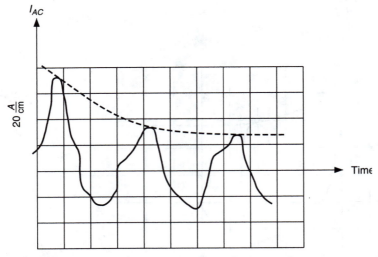

Figure 2.8 Transient response at:

$$I_{DC} = 14 \text{ amps}$$
$$I_{AC\backslash SS} = 36 \text{ amps}$$
$$V_{AC} = 57.5 \text{ volts}$$

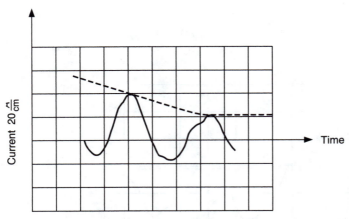

Figure 2.9 Transient response at $I_{DC} = 40$ amps $V_{AC} = 34$ volts

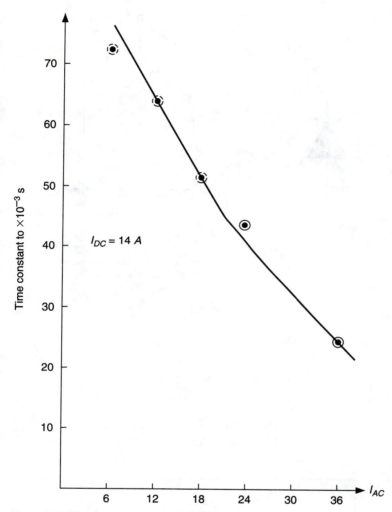

Figure 2.10   Time-constant versus AC Current.

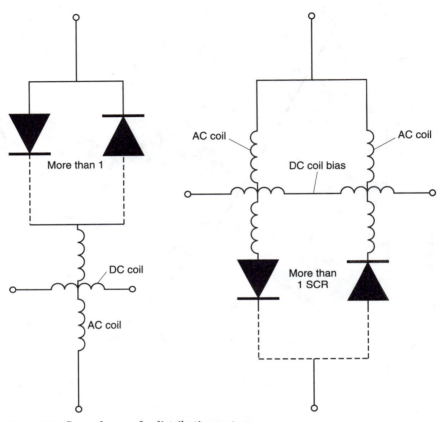

Figure 2.11 Surge damper for distribution system.

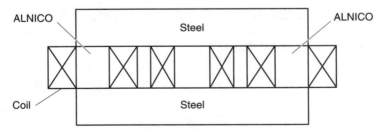

Figure 2.12 Cross section of saturistor.

This author may indicate that temperature in the saturistor rises at a fast rate such that some auxiliary means of heat dissipation process is needed in order to have an adequate sustained operation of the whole surge damper.

Presently, with aluminum straps, temperature rise is of the order of 1°C/sec. Therefore, more experimental tests, as well as theoretical analysis, are needed to control heat dissipation and the perfection of DC bias control on various levels of saturation, in connection with high-voltage SCRs. Now being developed are various kinds of ALNICO having different limits of coercive force with decreasing values of magnetic permeabilities. For example:

ALNICO 5 has coercive force of 600 oersteds.

ALNICO 9 has coercive force of 1600 oersteds.

2.3 Steady-State Characteristics

When excited with AC source, the saturistor becomes a resistor with a varying power factor, from a low value at small currents to a maximum at about twice the pickup current, with lower values at high currents due to saturation. The pickup current corresponding to the coercive force of the ALNICO is at the point at whic/h both resistance and reactance of the saturistor begin to rise rapidly with increase of current. Thus, the saturistor impedance increases with current and it forms a current-limiting device with low impedance at normal operating current. A saturistor may be used in the secondary circuit of an induction motor in order to provide a high resistance at starting and at low drives and a low resistance at full normal speed of operation without any control devices, as proposed in several recent IEEE papers.

The resistance of the saturistor is due to its hysteresis loss, which is proportional to the frequency, so the power factor is independent of frequency. Hence, the saturistor may be used to limit the initial current surge when the switch is closed on a lamp or electronic circuit.

In either case, one of the unresolved questions is what the magnitude and duration of the transient currents due to switching will be.

The time constant that governs the real decay of the current offset when the circuit is suddenly closed is L/R seconds, where L is the inductance and R is the resistance of the circuit. Does the resistance component due to hysteresis, which depends also on the current, decrease the time constant? Will the remainent magnetization of the ALNICO cause a large momentary current surge, as in a transformer? Or will the high power factor of the saturistor make the decrement rate so high that the peak current will be limited? And how can magnetic saturation be allowed for in calculating transient currents in a saturistor?

Figure 2.3 shows the B-H curve of ALNICO 5-7 used in the test. The coercive force is about 750 oersteds. Each of the three-phase leg has 50 turns, and the length of magnetic flux path in the ALNICO of each leg was 0.725 in. Therefore, the pickup current is:

$$I_p = 750(0.725)(2.02)(50) \div 2$$

$$= 15.5 \text{ amp rms}$$

The maximum permeability occurs at a coercive force of about the 1200 oersteds, where the peak flux density is about 1300 gauss, corresponding to a permeability of just over 10.

The initial permeability at very low currents is somewhat greater than that given by the slope of the recoil branch of the B-H curve, and is estimated to be about 2, from Fig. 2.3.

Figure 2.4 shows curves of resistance and reactance versus current found by tests at 60 cps, with the saturistor connected in Y. The resistance values shown are the measured total values, less 0.54 Ω, the value of the DC resistance of the winding.

The plotted resistance values are solely due to hysteresis and eddy-current losses. Also shown are plots of watts/in^3 of ALNICO versus current, which gives an indication of the rate of temperature rise.

The thickness of the ALNICO blocks, $^{0.3}\!/_{8}$ in, is small enough so that the 60-cps eddy-current losses are small.

The steel laminations in the yokes of the saturistor are $^1\!/_8$ in thick, so that the eddy-current losses in them are quite large at high flux densities.

The flux density in the ALNICO is limited by saturation to about 14,000 gauss, which is reached when the rms current is about twice the pickup current, or about 30 amperes. The increase in losses at higher currents is assumed to be almost all due to eddy-current losses in the steel.

However, the eddy-current losses in the ALNICO also increase at higher currents with the same peak flux, because the flux density rises more rapidly in the first part of each cycle, creating third and higher harmonic components in the flux wave.

Figure 2.5 shows plots of R, X, Z of a Δ-connected saturable-resistor, in addition to watts/in^3 versus phase current.

As expected, the reactance and resistance values at higher currents are being reduced sharply due to the effects of third harmonics.

In summary, we can indicate:

1. The steady-state Δ connection of a three-phase saturable-resistor signified the effects of the third harmonic presence on the resistance and reactance values at high currents. This effect will be reflected on its power factor and the time-constant.

2. The variation of watts/in^3 in the ALNICO is a measure of the rate of its temperature rise, and this can be used in predicting the temperature inside this core material.

2.4   Transient Performance

The three-phase saturistor was connected with one phase in series with the other two in parallel, as shown in Fig. 2.13. This corresponds to the instant on three-phase when the current in phase A is a maximum, and the other two phases carry half current in the opposite direction.

With the coils connected in this way, a direct current was passed through first to fully magnetize leg A, and then an AC voltage was applied suddenly.

Figures 2.14 and 2.15 show transient response for AC voltages applied suddenly after two different levels of DC premagnetization.

Those figures show the strength of the current offset, and also a prediction of the rate of current decay toward its steady-state value.

The transient test reveals that the peek offset depends on the level of premagnetization and also on the phase angle of the AC voltage applied at the time of closing the switch. Another factor is the peak value of AC voltage when the surge current is below or above the pickup value, which, in effect, gives different values of time-constant L/R.

Figures 2.14 and 2.15 show different offset values of transient currents, each with a different time-constant.

Transient performance of such type of reactor is very different from any other inductive circuit. Much more can be explored about such properties in a variety of multiphase connections.

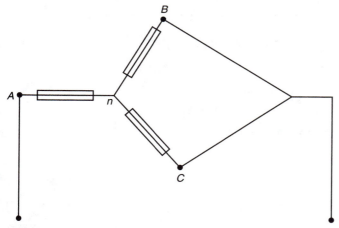

Figure 2.13 Unbalanced three-phase saturistor.

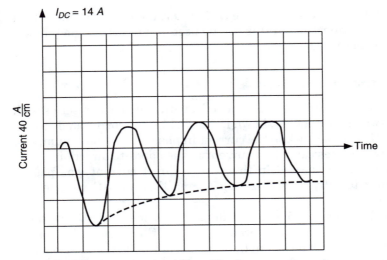

Figure 2.14   Transient response at $V_{AC} = 30$ volt.

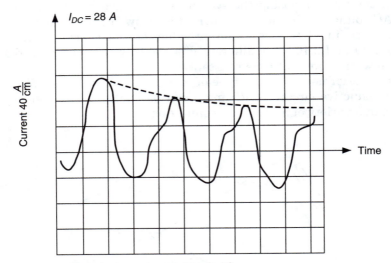

Figure 2.15   Transient response at 43-volt AC.

Transient tests have been carried out by varying AC voltages super-imposed on different levels of DC premagnetization.

The tests revealed the changeable effect of the time-constant on the current offset values in each case, subject to the phase angle of the applied voltage and residual DC flux.

Such a study will have an important reflection on the control of surge current in domestic and industrial applications.

2.5 Temperature Effects

Steady-state tests have been carried out on a three-phase saturable-resistor for Δ configuration, and then on Y-connection for the determination of phase resistance, reactance, impedance, and power-loss density in watts/inch^3. The first cycle of experimental tests started with the saturistor at ambient temperature of about 25°C. Immediately afterward, with the device hot at an absorbed power density indicated, a new, second cycle of tests was repeated as in cycle one. Figures 2.16 through 2.23 show the trend for changes in the resistance, reactance, impedance, and the power-loss density in W/in^3.

Again, looking at Figs. 2.16 and 2.23 for the phase resistance for Δ and Y-connections, the presence of third harmonic currents Δ in the device played a major role in elevating the resistance beyond that at a line current of around 25 amp. In the Y device, uniformity had been observed throughout the range of line current.

Turning again to Figs. 2.18 and 2.22, which display plots for the phase impedance for the Δ and the Y configuration, again the uniformity is seen for some reduction in reactance over a line current range of up to about 30 amp, and the uniform increase for a lone current, beyond the 30 amp. For the Δ connection, the phase reactance at hot condition displayed an increase at a line current below that at 20 amp and beyond 31 amp (approximately). The reactance at hot condition for a line current ranging between 20 and 31 amp showed a remarkable reduction, indicating a very interesting phenomenon.

The impedance pattern for the Δ connection indicated by Fig. 2.17 demonstrates a reverse behavior from that of its reactance at the same time, while keeping uniformity. For Y, the connection as shown in Fig. 2.21 the pattern is again the reverse behavior for that of its reactance at the same time while keeping uniformity. For the connection, the pattern is again the reverse behavior of its reactance, whereby at hot condition the phase impedance shows a reduction in ohmic values up to a line current of about 20 amp and beyond 40 amp, while the impedance was higher at a line current range between 20 and 40 amp. All of these and verse changes could be attributed to the severe nonlinear character of the saturistor and effects of circulating harmonic in the case of Δ connection.

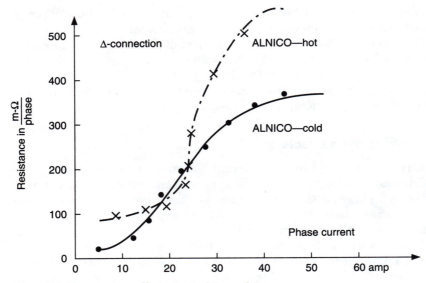

Figure 2.16 Temperature effect on saturistor resistance.

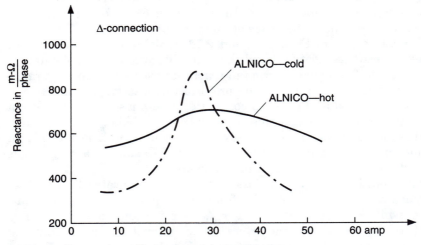

Figure 2.17 Temperature effect on saturistor reactance.

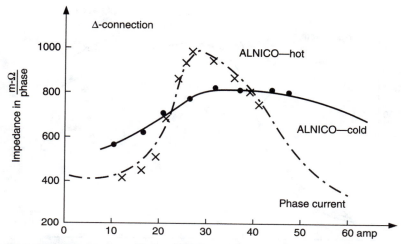

Figure 2.18 Temperature effect on saturistor impedance.

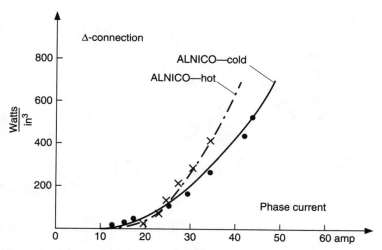

Figure 2.19 Power loss density in saturistor.

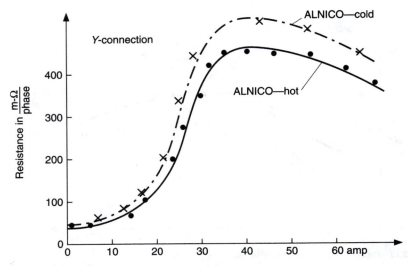

Figure 2.20 Temperature effect on saturistor resistance.

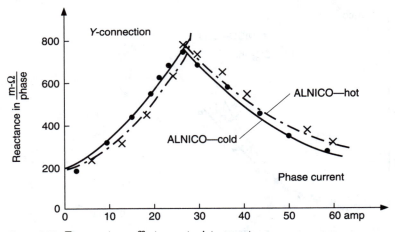

Figure 2.21 Temperature effect on saturistor reactance.

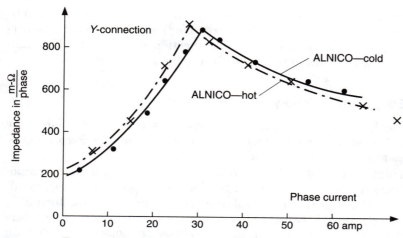

Figure 2.22 Temperature effect on saturistor impedance.

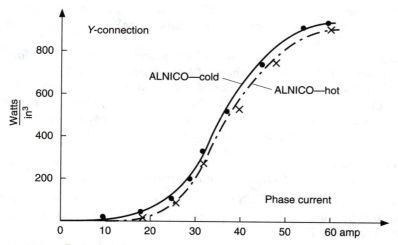

Figure 2.23 Power loss density in saturistor.

Pattern for power-loss density in watts/in^3 for the Δ connection indicated by Fig. 2.19 shows a uniform trend for increase for the hot saturistor, while the pattern for the Y connection indicated by Fig. 2.23 demonstrates a uniform reduction for the hot saturistor.

2.6   Eddy-Current Theory

Analytical, as well as empirical, formulae have been established for calculating eddy-current losses for ordinary ferromagnetic core material used in saturable-reactors.

Calculation of induced electromagnetic steady-state field components for saturable-resistor must be given an important priority, leading up to the objective of calculating eddy-current losses. Including electromagnetic field calculations in regular ferromagnetic of high permeability has been carried out extensively by previous authors for the induced electric field, the position of the separating plane, and the surface impedance, as well as closed-form solutions for eddy-current losses based on the integral form of Maxwell equations.

However, the peculiar properties of the saturable-resistor, as indicated earlier, require the formulation procedural calculations for induced fields as well as power losses.

In the following subsections, the author intends to develop the following: Assuming that the thickness of half of the core block is much greater than the depth of penetration of the separating plane (the separating plane is a measure through which the rate of change of magnetization advances inward), the tasks outlined earlier will be analyzed for two distinct cases. These are:

1. The electrical conductivity σ is assumed constant and independent of changes of current, frequency, and temperature.

2. The electrical conductivity is a function of the line current (an actual case), and hence of the magnetic field before and while saturation occurs. (See App. B.

Experimental verification is given for the theoretical formulae developed for both of these cases.

2.6.1   Solution of the induced electric field and the separation plane

Constant electrical conductivity.   The value of σ is taken to correspond to the regime where the saturable-resistor possesses its maximum resistance.

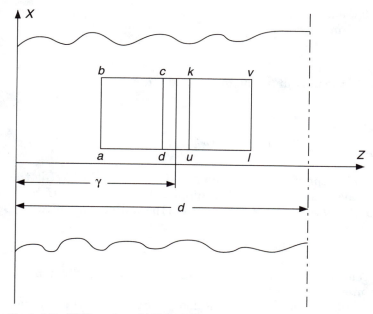

Figure 2.24 (*X-Z*) section of ALNICO block.

Refer to Fig. 2.2 and let

σ = electrical conductivity

e = electric field polarized in the x direction

In Fig. 2.2, h and b show magnetic field intensity and induction in the y direction. Maxwell's integral equations are:

$$\oint \bar{h} \cdot d\bar{s} = \sigma \int_s \bar{e} \cdot d\bar{a} \qquad (2.1)$$

$$\oint \bar{e} \cdot d\bar{s} = - -\frac{\delta}{\delta t} \int_s \bar{b} \cdot d\bar{a} \qquad (2.2)$$

where s denotes surface.

Equations (2.1) and (2.2) become:

$$\oint \bar{u}_y h \cdot ds = \delta \int e \cdot da \qquad (2.3)$$

$$\oint u_x e \cdot ds = -\frac{\delta}{\delta t} \int b \cdot da = -\frac{\delta \phi}{\delta t} \qquad (2.4)$$

where 0 = magnetic flux
             u_x, u_y = unit vectors in the x and v directions

$$h = H_m \sin \omega t \tag{2.5}$$

H_m = the maximum magnetic field intensity

and

$$b = (B_0 Sig H) U(h - H_c) \tag{2.6}$$

where $U(h - H_c)$ represents a delayed step function and Sig H is the unit square wave associated with the applied sinusoidal wave H.

The rate of change of flux per unit length in the x direction is given by:

$$\frac{d\phi}{dt} = 2vB_0 \tag{2.7}$$

where v = velocity of travel of the position of the separating plane toward the interior of the core

Refer to Fig. 2.24 and consider the loop $(lvku)$. Since there is not change of magnetic flux there, no induced field beyond the position of the separating plane y will exit.

Also considering the loop $(ukcd)$ enclosing the plane y; the voltage equations can be written as

$$e_{uk} + e_{cd} = 2B_0 \frac{dy}{dt} = 2vB_0 \tag{2.8}$$

since $e_{uk} = 0$. Therefore

$$e_{cd} = 2B_0 \frac{dy}{dt} = E \tag{2.9}$$

where y represents the advance of the position of the separating plane in the z direction.

Now refer to Fig. 2.25 to calculate the value of the integral in Eq. (2.3); take loop (pqkl) where $h = H_c$ at the starting position of the separating plane. Therefore

$$\oint_{pqkl} u_y h \cdot ds = H = H_c = \sigma y E \tag{2.10}$$

or

$$\frac{\delta h}{\delta z} = \frac{H = H_c}{y} = \sigma E \tag{2.11}$$

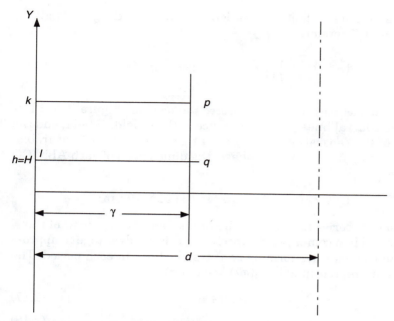

Figure 2.25 (Y-Z) section of ALNICO block.

where H = the value of the field h at $z = 0$

Therefore, the field h is linearly diminished from its initial value at $z = 0$ up to a point where h will again regain the value of H_c at which the separating plane terminates.

From Eqs. (2.9) and (2.11), one can write:

$$H - H_c = 2\sigma y B_0 = \sigma B_0 \frac{dy^2}{dt} \tag{2.12}$$

Substituting for H and H_c from Eq. (2.5):

$$H_m sin\omega t - H_m sin\omega t_c = \sigma B_0 \frac{dy^2}{dt} \tag{2.13}$$

Therefore, with $y = 0$ and $t = t_c$ the solution of y is:

$$y = U_{-1}(t - t_c)\left(\frac{\sqrt{H_m}}{\sigma\omega B_0}\right)\left(\sqrt{cos\omega t_c - cos\omega t + (\omega t_c - \omega t)sin\omega t_c}\right) \tag{2.14}$$

Then, substituting in Eq. (2.9) for y will result in a solution for the E field

$$E = U_{-1}(t - t_c)\left(\frac{H_m B_0 \omega}{\sigma}\right)\left(\frac{sin\omega t - sin\omega t_c}{\sqrt{cos\omega t_c - cos\omega t + (\omega t_c - \omega t)sin\omega t_c}}\right) \tag{2.15}$$

The maximum depth of penetration is found by setting in Eq. (2.14) $\omega t = (\pi - \omega t_c)$. Therefore,

$$\delta = \left(\frac{\sqrt{H_m}}{\sigma \omega B_0} \right) \left(\sqrt{2 cos \omega t_c} + (2 \omega t_c - sin \omega t_c) \right)^{\frac{1}{2}} \tag{2.16}$$

The fundamental component of the induced electric field. Figure 2.26 illustrates the general linearized appearance of the E field, H field, and the position of the separating plane y for a three-phase, 5-kvA saturable-resistor having five blocks per phase; the dimensions of each ALNICO block are:

$$(1.84 \times 10^{-2})(2.54 \times 10^{-2} \text{ m})(0.952 \times 10^{-2} \text{ m})$$

Using Fourier series to calculate the fundamental component of the E field resulted in a complicated expression. It is of advantage to approximate the voltage wave form by two straight lines in each period. The two lines are represented by $e_1(\omega t)$ and $e_2(\omega t)$.

$$e = a\theta + B \tag{2.17}$$

$$\theta = \omega t, \ \theta_c = \omega t_c \tag{2.18}$$

$$a = \frac{E_m}{\theta_m - \theta_c}, \ B = \frac{E_m \theta_c}{\theta_c - \theta_m} \tag{2.19}$$

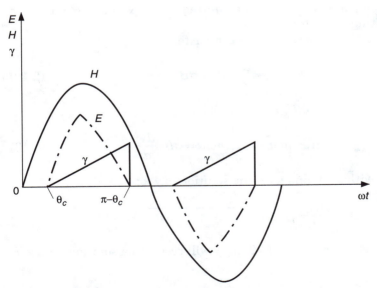

Figure 2.26 Waveforms of E, H, and γ below saturation (constant conductivity).

$$e_2 = A\theta + B \tag{2.20}$$

$$A = \frac{E_m}{(\theta_m + \theta_c - \pi)}, \quad b = \frac{E_m(\theta_c - \pi)}{(\theta_m + \theta_c - \pi)} \tag{2.21}$$

The fundamental component of E is then:

$$E_f = U(t - t_c) \sqrt{\frac{H_m B_0 \omega}{\sigma}} \left[(K_1 + K_1')cos\omega t + (K_2 + K_2')sin\omega t \right] \tag{2.22}$$

The expressions for K_1, K_2, K_1', and K_2' are in App. A. The maximum of E_f is.

$$|E_f max| = \sqrt{\frac{H_m B_0 \omega}{\sigma}} \left[(K_1 + K_1') + j(K_2 + K_2') \right] \tag{2.23}$$

The surface impedance and power loss density.   The field h is sinusoidal at $z = 0$ and decreases linearly inside the core.

$$H - H_m sin\omega t \tag{2.24}$$

Let Z be the complex surface impedance

$$Z = r + JX$$

A is well known. It can be expressed as:

$$A = \frac{\text{Max of fundamental of } E}{\text{Max of fundamental of } H} \tag{2.25}$$

The poynting vector $\overline{W}$ into the core is:

$$\overline{W} = U + jv$$

$$= \frac{1}{2} E_m H_m{}^* \tag{2.26}$$

From Eqs. (2.25) and (2.26)

$$U = R_e \overline{W} = \frac{1}{2} rH_m^2 W/m^2 \tag{2.27}$$

and

$$A = \frac{\sqrt{\dfrac{H_m B\omega}{\sigma}} \left[(K_1 + K_1') + J(K_2 - K_2') \right]}{-jH_m} \tag{2.28}$$

denoting now:

$$R = -\sqrt{\frac{\omega B_0}{\sigma H_m}} \, (K_2 + K_2') \tag{2.29}$$

$$X = \sqrt{\frac{\omega B_0}{\sigma H_m}} \, (K_1 + K_1') \tag{2.30}$$

and

$$\alpha = tan^{-1} \frac{(K_1 + K_1')}{(K_2 + K_2')} \tag{2.31}$$

as the resistance, reactance, and phase angle of the surface impedance, respectively.

Variable electrical conductivity. The resistive component of the impedance of the saturable-resistor changes with the AC current. The corresponding electrical conductivity varies in the two distinct regions, below saturation and at saturation, respectively.
The conductivity in region 1 is:

$$\sigma = \frac{1}{mH + n} \tag{2.32}$$

and in region 2 is:

$$\sigma_2 = \frac{1}{m'H + n'} \tag{2.33}$$

Where m, n, m', and n' are constants as shown in App. 8.
Using an approach similar to case 1, the following solutions are obtained for the position of the separating plane, induced electric field, surface impedance, and eddy-current power-loss density:

$$\gamma = \frac{1}{\sqrt{B_0}} \left[\omega t \left(\frac{k_1}{2\omega} - \frac{k_3}{\omega} \right) - \left(\frac{k_1}{4\omega} \, sin2\omega t + \frac{k_2}{\omega} \, cos\omega t \right) \right] \tag{2.34}$$

$$E = \sqrt{B_0} \left\{ \frac{k_2 sint - \frac{k_1}{2} cos2t + \frac{k_1}{2} - k_3}{\left[\omega t \left[\left(\frac{k_1}{2\omega} - \frac{k_3}{\omega} \right) - \left(\frac{k_1}{4\omega} \, sin2\omega t + \frac{k_2}{\omega} \, cos\omega t \right) \right] \right]^{\frac{1}{2}}} \right\} \tag{2.35}$$

The depth of penetration is obtained by setting $\omega t - (\pi - \omega t_c)$ in Eq. (2.34).

$$\delta = \frac{1}{\sqrt{B_0}} \left\{ (\pi - \omega t_c)\left(\frac{k_1}{2\omega} - \frac{k_3}{\omega}\right) - \left(\frac{k_1}{4\omega} sin2\omega t_c - \frac{k_2}{\omega} cos\omega t_c\right) \right\}^{\frac{1}{2}} \quad (2.36)$$

where

$$k_1 = kH_m^2 \quad (2.37)$$

$$k_2 = nH_m - kH_cH_m \quad (2.38)$$

$$k_3 = 1NH_c \quad (2.39)$$

$$k = \frac{ml}{N} \quad (2.40)$$

H_c = coercive force in A-T/m
H_m = peak magnetic field intensity in A-T/m
l = length of core in the direction of magnetization, in meters
N = number of turns per phase
m, n = constants in the electrical conductivity equation before saturation; m', n' are constants in the saturation region

$$Z = r + jx = \frac{E_{Fmax}}{H_{max}} \quad (2.41)$$

The eddy-current loss density is:

$$W = 1/2rH_m^2 \quad (2.42)$$

A closed-form calculation of the fundamental of the electric field in Eq. (2.35) by Fourier series expansion is too complicated. Therefore, striving toward a physically meaningful conclusion, the wave forms of E and γ are plotted in Figs. 2.27 and 2.28 for a specific three-phase saturable-resistor. The cores consist of five ALNICO blocks:

$$(1.84 \times 10^{-2} \text{ m})(2.54 \times 10^{-2} \text{ m})(0.952 \times 10^{-2} \text{ m}) \quad (2.43)$$

The other magnetic data are:

l = 1.84×10^{-2} m in the direction of magnetization
a = 2.42×10^{-4} m^2
μ = 2.5
H_c = 6×104. A-T/m as the coercive force
H_m = 9.85×104 A-T/m as the peak impressed magnetomotive
B_0 = 0.188 tesla

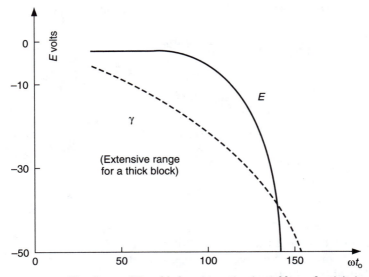

Figure 2.27 Waveforms of E-and-below saturation (variable conductivity).

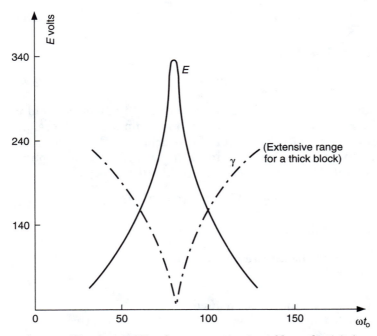

Figure 2.28 Waveforms of E and γ at saturation (variable conductivity).

Plots of the electric field and of the position of the separating plane shown in Fig. 2.27 are based on the actual dependence of the electrical conductivity on the line current and the magnetic field below saturation. Figure 2.28 corresponds to the plots at saturation.

The fundamental component found by Fourier analysis is:

$$E_F = |E_F| sin\omega t \qquad (2.44)$$

$$|E_F| \frac{1}{\pi} \int_{\omega tc}^{\pi - \omega t} E(\omega t) sin\omega t d\omega t \qquad (2.45)$$

$E_F(\omega\tau)$ is almost a constant amplitude approximately 46 volts. The calculated amplitude of the fundamental is:

$$E_{F_{max}} = 22 \text{ V}$$

2.6.2 Experimental verification of the theory

Calculation of power-loss density at constant. Referring to Eqs. (2.21) and (2.24) and assuming a constant value of 0.076 $\omega - m$ (corresponds to the maximum resistive component of the saturable-resistor and a peak value of impressed magnetic field of 9.85×10^4 A-T/m and coercive force of 6×10^4 A-T/m), the eddy-current density was calculated to be 126×10^4 W/m^2 and 0.5122W/m surface resistance.

Calculation of power-loss density at variable. This is confined to the case of the maximum resistive component for the saturable-resistor. Referring to Eqs. (2.33), (2.36), and (2.37), at the same values for peak impressed magnetic field and coercive force, as in the eddy-current power-loss density was calculated to be 93×10^4 W/m^2, corresponding to a surface resistance of 0.347 Ω/m.

Experimental work. Tests were conducted on the same saturable-resistor, applying a series of values of AC voltages and measuring wattmeter as well as current readings. After deducting the DC cooper loss from the wattmeter readings and the DC resistance from the resistive component of the saturable-resistor, the total iron loss per cubic meter and the net resistance per phase are obtained and plotted with respect to line current in Fig. 2.1).

In Fig. 2.29, it is indicated that the maximum value of net resistive component occurs at 26.5 A with a corresponding value of 19.52×10^6 W/m^2, compare to 93×10^4 W/m^2 calculated in case 2.

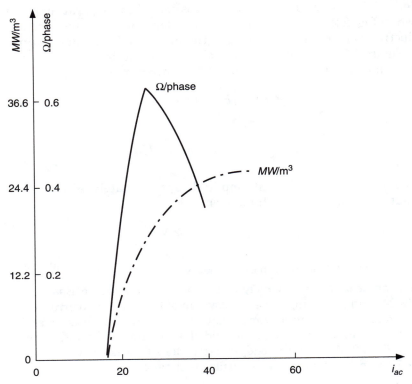

Figure 2.29 Experimental curve for the resistive component and iron losses of a three-phase saturable resistor against line current.

2.6.3 Remarks

1. Closed-forms theoretical formulas based on the integral form of Maxwell equations have been developed to calculate eddy-current losses for thick laminations of hard ferromagnetic material core in saturable-resistors. In the first case the established expressions are based on a constant, independent electrical conductivity for the saturable-resistor. In the second case an actual distinct functional dependence on the AC line current before and through saturation have been considered.

2. Experimental tests conducted on a three-phase saturable-resistor having ALNICO 5-7 cores indicated that eddy-current theoretical formula based on actual dependence of the electrical conductivity on AC line current gave a result about 9 percent higher than the measured value. This percentage should be slightly higher because hysteresis losses are neglected in the (0.952×10^{-2})-mA-thick

ALNICO blocks used. It had been established that in thick laminations, eddy-current losses are proportional to large changes of current
amplitudes. Eddy-current losses as calculated with constant electrical
conductivity are substantially higher than these obtained from tests.

3. Solutions for the induced electric field, the position of the separating plane (representing advance of the change of magnetization and
the depth of penetration), have been established in the region confined
between the initial pickup of magnetization and the maximum resistance, and then at saturation. Plots indicate that in a typical saturable-
resistor, the electric field has an almost odd rectangular waveshape
before saturation (see Fig. 2.33). Symmetrically peaked waveforms
result throughout the saturation region. The position of the separating
plane indicates a pattern of linear decrease before saturation. The
position of the plane is different thereafter. At saturation, the position
of the separating plane shows two trends: first, a steep decrease in the
first quarter cycle and second, a steep rise in the following period. (See
Fig. 2.34)

4. Further work is required to generalize the eddy-current formulas
developed in this book. It has been established that the electrical conductivity depends on changes in temperature as well as frequency. Even
with sinusoidal applied voltage, the magnetizing current waveshape will
be a distorted sinusoid, introducing strong triple harmonics. Therefore,
the next step in this analysis is to develop formulas for eddy-current loss
taking into account the effects of temperature and harmonics.

2.6.4 Appendix A

Following are the Fourier coefficients of the induced electrical field
waveform (constant conductivity).

The amplitudes of the fundamental components of e_1 are:

$$k_1 = \frac{1}{\pi} \int_{\theta c}^{\theta m} (a\theta + b)\cos\theta d\theta \tag{A.1}$$

$$k_2 = \frac{1}{\pi} \int_{\theta c}^{\theta m} (a\theta + b)\sin\theta d\theta \tag{A.2}$$

Then the amplitudes of the fundamental components of e_2 are:

$$k_1' = \frac{1}{\pi} \int_{\theta m}^{\pi - \theta c} (A\theta + B)\cos\theta d\theta \tag{A.3}$$

$$k_2 = \frac{1}{\pi} \int_{\theta m}^{\pi - \theta c} (A\theta + B)\sin\theta d\theta \tag{A.4}$$

$$k_1 = \frac{1}{\pi} \left[a\cos\theta_m - \cos\theta_c + \theta_m \sin\theta_m - \theta_m \sin\theta_c + b\sin\theta_m - \sin\theta_c \right] \quad \text{(A.5)}$$

$$k_2 = \frac{1}{\pi} \left[a\sin\theta_m - \sin\theta_c + \theta_c \cos\theta_c - \theta_m \cos\theta_m + b\cos\theta_c - \cos\theta_m \right] \quad \text{(A.6)}$$

$$k_1' = \frac{1}{\pi} \left[A\pi\sin\theta_c - \theta_c \sin\theta - \cos\theta_m - \cos\theta_c - \theta_m \sin\theta_m + B\sin\theta_c - \sin\theta_m \right]$$
$$\text{(A.7)}$$

$$k_2' = \frac{1}{\pi} \left[A\sin\theta_c \sin\theta_m - \theta_c \cos\theta_c - \theta_m \cos\theta_m + \pi\cos\theta_c + B\cos\theta_m + \cos\theta_c \right]$$
$$\text{(A.8)}$$

where a, b, A, B are constants related to the linear functions of induced electric field that can be determined according to Eqs. (2.18) and (2.19) in the text.

2.6.5 Appendix B

The electrical resistivity (see Fig. 2.30) is as follows:

1. In the region between pickup and peak resistance (below saturation)

$$\rho = (ai + b) \qquad \text{(B.1)}$$

$$\sigma = \frac{1}{(ai - b)} \qquad \text{(B.2)}$$

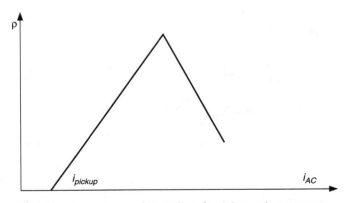

Figure 2.30 Approximate electrical conductivity against current.

$$H = \frac{N_i}{l} \tag{B.3}$$

$$\sigma = \frac{1}{\dfrac{al}{N} H + b} = \frac{1}{mH + n} \tag{B.4}$$

The value of the resister used in the texts is:

$$\sigma = \frac{1}{4.36 \times 10^{-4}H - 18.91} \; \Omega - m \tag{B.5}$$

2. In the region beyond peak resistance (saturation)

$$\rho = (a'i + b) \tag{B.6}$$

$$\sigma = \frac{1}{a'i} + b \tag{B.7}$$

or

$$\sigma = \frac{1}{\dfrac{a'l}{N} + b} = \frac{1}{m'H + n} \tag{B.8}$$

$$\sigma = \frac{1}{-6.23 \times 10^{-3}H + 819.52} \; \Omega - m \tag{B.9}$$

where H is in At/m.

2.7   Saturistor with Actual *B-H* Curve

Figure 2.31 shows the actual displaced $B\text{-}H$ magnetization curve of the hard ferromagnetic core, where the magnetic flux density B follows a geometrical power series in its buildup pattern, and the displaced magnetic field intensity H increases according to an arithmetical series. Figure 2.32 illustrates a block of hard magnetic core of thickness $2\,d$. X-polarized plane waves impinge on both faces, propagating toward the core center.

It is assumed that $d \gg \delta$, where δ is the depth of the magnetic field penetration.

Solutions are required for the induced electric field, the surface impedance, the separating plane, and the eddy-current loss power density. These objectives are based on two cases with respect to the core electrical conductivity, namely:

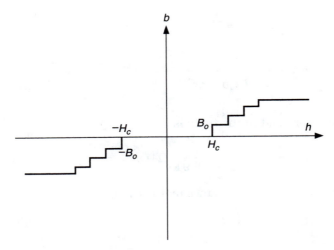

Figure 2.31 Magnetization curve.

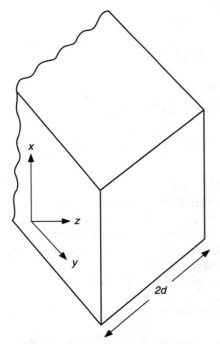

Figure 2.32 Field incidence at the surface of ALNICO.

1. σ is independent and constant.

2. σ is a function of the applied mmf below saturation.

Experimental work has been conducted to test the validity of the theoretical formulae developed.

At constant electrical conductivity

where σ = electrical conductivity

e = electric field polarized in the X direction

$h \cdot b$ = magnetic field intensity and induction in the y direction

Maxwell's integral equations are:

$$\oint \bar{h} \cdot d\bar{s} = \sigma \int_s \bar{e} \cdot d\bar{a} \qquad (2.46)$$

$$\oint \bar{e} \cdot d\bar{s} = \frac{\delta}{\delta t} \int_s \bar{b} \cdot d\bar{a} \qquad (2.47)$$

where s denotes surface.

Equations (2.46) and (2.47) may be put into the following form:

$$\oint \hat{J}h \cdot ds = \sigma \int \bar{e} \cdot d\bar{a} \qquad (2.48)$$

$$\oint \hat{i}e \cdot ds = \frac{\delta}{\delta t} \int b \cdot da = -\frac{\delta \psi}{dt} \qquad (2.49)$$

where ψ = total magnetic flux

$\hat{i}, \hat{J}$ = unit vectors in the X and y directions

$$h = H_m sin\omega t \qquad (2.50)$$

H_m = the applied magnetic field intensity

and

$$b = B.Sig \sum_0^n \frac{1 - \rho^{n+1}}{1 - \rho} U[h - H_e(1 + nk)] \qquad (2.51)$$

where $Sig \sum_0^n \dfrac{1 - \rho^{n+1}}{1 - \rho}$

refers to a multisquare-wave buildup associated with the applied field H.

where ρ = the base of the geometrical series

$$= \frac{\Delta b}{B_0}$$

B_0 = the initial step increase in the b field immediately after the field h surpassed the coercive field H_c

$Uh - H_c(1 + nk)$ represents a delayed step function

$k = \dfrac{\Delta h}{H_c}$ the base of the arithmetical series for the increase of

the field h

The rate of change of magnetic flux per unit length in the X direction is given by:

$$\frac{d\omega}{dt} = 2vB_0 \sum_{0}^{n} \frac{1 - \rho^{n+1}}{1 - \rho} \tag{2.52}$$

where v = velocity of propagation of the separating plane toward the interior of the core.

Refer to Fig. 2.24 and consider the loop (lvku). Because there is no change of magnetic flux, no induced field beyond the separating plane τ will exist.

Also with respect to the loop (ukcd) enclosing the plane τ, the voltage equations can be expressed as

$$e_{ps} + e_{fb} = 2B_0 \sum_{0}^{n} \frac{1 - \rho^{n+1}}{1 - \rho} \frac{d\tau}{dt} = 2vB_0 \sum_{0}^{n} \frac{1 - \rho^{n+1}}{1 - \rho} \tag{2.53}$$

since $e_{ps} = 0$, therefore,

$$e_{fb} = 2 \sum_{0}^{n} \frac{1 - \rho^{n+1}}{1 - \rho} \frac{d\tau}{dt} = E \tag{2.54}$$

τ represents the propagation of the location of the separating plane in the z direction.

Refer to Fig. 2.25 for calculating the value of the integral in Eq. (2.3). Take the loop (qlkp): $h = H_c$ at the initial location of the separating plane

$$\oint Jh \cdot ds = H - H_c = \sigma rE \tag{2.55}$$

or

$$\frac{\partial h}{\partial Z} = \frac{H - H_c}{\tau} = \sigma E \tag{2.56}$$

where H = the field intensity h at $Z = 0$.

From Eqs. (2.54) and (2.56), one can write:

$$H - H_c = 2\sigma\tau \sum_0^n \frac{1 - \rho^{n+1}}{1 - \rho}$$

$$= \sigma B_0 \sum_0^n \frac{1 - \rho^{n+1}}{1 - \rho} \frac{d\tau^2}{dt} \qquad (2.57)$$

Then, since $r = 0$ at $t = t_c$, the solution for r is:

$$\gamma = u(t - tc) \left[\frac{H_m[cos\omega t_c - cos\omega t + (\omega t_c - \omega t)sin\omega t_c]}{\sigma\omega B_0 \sum_0^n \frac{1 - \rho^{n+1}}{1 - \rho}} \right]^{\frac{1}{2}} \qquad (2.58)$$

Substituting γ in Eq. (2.54) results in:

$$E = u(t - tc) \left(\frac{H_m\omega B_0 \sum_0^n \frac{1 - \rho^{n+1}}{1 - \rho}}{\sigma} \right)$$

$$\left(\frac{sin\omega t - sin\omega t_c}{\sqrt{cos\omega t_c - cos\omega t + (\omega t_c - \omega t)sin\omega t_c}} \right) \qquad (2.59)$$

Then, from Eq. (2.58), the maximum depth of penetration δ can be obtained by inserting $\omega t = (\pi - \omega t_c)$.

$$\delta = \left[\frac{H_m}{\sigma\omega B_0 \sum \frac{1 - \rho^{n+1}}{1 - \rho}} \right] [2cos\omega t_c + (2\omega t_c - \pi)sin\omega t_c] \qquad (2.60)$$

Proceed to calculate the fundamental component for the induced electric field. Fourier series expansion has been applied on a straight-line approximation on the field waveform representing the behavior on a 5-KVA reactor with hard ferromagnetic core.

The fundamental component E_f:

$$E_f = U(t - t_c) \left\{ \frac{\omega H_m B_0 \sum_0^n \frac{1 - \rho^{n+1}}{1 - \rho}}{\sigma} \right.$$

$$\left. \left[\left(N_1 + N_1'\right) cos\omega t + \left(N_2 + N_2'\right) sin\omega t \right] \right\} \qquad (2.61)$$

where

$$N_1 = \frac{1}{\pi}[a\,(cos\theta_m - cos\theta_c + \theta_m\,sin\,\theta_m - \theta_c\,sin\theta_c) + b\,(sin\theta_m - sin\theta_c)] \quad (2.62)$$

$$N_2 = \frac{1}{\pi}[a\,(sin\,\theta_m - sin\theta_c + \theta_c\,cos\theta_c - \theta_m\,cos\theta_m) + b(cos\theta_c - cos\theta_m)] \quad (2.63)$$

$$N_1' = \frac{1}{\pi}[A(\pi\,sin\theta_c - \theta_c\,sin\theta_c - cos\theta_m - cos\theta_c - \theta_m\,sin\theta_m) + B(sin\theta_c - \theta_m)]$$
$$(2.64)$$

$$N_2' = \frac{1}{\pi}[A(sin\,\theta_c - sin\theta_m - \theta_c - cos\theta_c + cos\theta_m + \pi\,cos\theta_c) + B\,(cos\theta_m + cos\theta_c)]$$
$$(2.65)$$

a, b, A, and B are constants relevant to the approximate linear function of the field E.

$$E_{fmax} = \left[\frac{\omega H_m B_0 \Sigma_0^n \frac{1 - \rho^{n+1}}{1 - \rho}}{\sigma}\right]^{\frac{1}{2}} \left[(N_1 + N_1') + J(N_2 + N_2')\right] \quad (2.66)$$

The surface impedance $Z = r + jx$ can be expressed as:

$$Z = \frac{E_{fmax}}{H_{max}} \quad (2.67)$$

and the power loss density vector

$$\bar{P} = U + jV \quad (2.68)$$

$$= \frac{1}{2}E_{fmax} \cdot H_{max} \quad (2.69)$$

$$U = R_e\,\bar{P} = \frac{1}{2}rH_{max}^2 \text{ watts/m}^2 \quad (2.70)$$

$$Z = \frac{\frac{\omega B_0 H_m}{\sigma}\,\Sigma_0^n \frac{1 - \rho^{n+1}}{1 - \rho}\left[(N_1 + N_1') + J(N_2 + N_2')\right]}{-jH_m} \quad (2.71)$$

where

$$r = -\left[\frac{\omega B_0 \sum_0^n \dfrac{1 - \rho^{n+1}}{1 - \rho}}{\sigma H_{max}}\right]^{\frac{1}{2}} \left(N_2 + N_2'\right) \tag{2.72}$$

$$x = -\left[\frac{\omega B_0 \sum_0^n \dfrac{1 - \rho^{n+1}}{1 - \rho}}{\sigma H_{max}}\right]^{\frac{1}{2}} \tag{2.73}$$

r, x are the surface resistance and reactance, respectively.

Electrical conductivity as an inverse linear function of the applied mmf. The electrical resistivity has two distinct regions of variations. The first is below the peak value and the other beyond that limit.

$$\sigma \text{ in the first region} = \frac{1}{mh + n} \tag{2.74}$$

$$\sigma \text{ in the second region} = \frac{1}{m'h + n'} \tag{2.75}$$

where n, nm', and n' are constants which depend on the particular kind of hard magnetic core. Using an identical approach compatible to that of case 1, the following solution was obtained for the propagation of the separating plane, the induced electric field, and the depth of penetration:

$$\tau = \frac{1}{\sqrt{B_0 \sum_0^n \dfrac{1 - \rho^{n+1}}{1 - \rho}}} \left[\omega t\left(\frac{k_1}{2\omega} - \frac{k_3}{\omega}\right) - \left(\frac{k_1}{4\omega} sin2\omega t + \frac{k_2}{\omega} cos\omega t\right)\right]$$

$$\tag{2.76}$$

$$E = \sqrt{B_0 \sum_0^n \frac{1 - \rho^{n+1}}{1 - \rho}} \left[\frac{\left(k_2 sin\omega t - \dfrac{k_1}{2} cos2\omega t + \dfrac{k_1}{2} - k_3\right)}{\omega t\left(\dfrac{k_1}{2\omega} - \dfrac{k_3}{\omega}\right) - \left(\dfrac{k_1}{4\omega} sin2\omega t + \dfrac{k_2}{\omega} cos\omega t\right)}\right]^{\frac{1}{2}} \tag{2.77}$$

and the depth of penetration δ is:

$$\delta = \frac{1}{B_0 \sum_0^n \dfrac{1 - \rho^{n+1}}{1 - \rho}} \left[(\pi - \omega t_c)\left(\frac{k_1}{2\omega} - \frac{k_3}{\omega}\right) - \left(\frac{k_1}{4} \omega sin2\omega t_c + \frac{k_2}{\omega} cos\omega t_c\right)\right]$$

$$\tag{2.78}$$

where $k_1 = kH_m^2$ (2.79)
$$k_2 = nH_m - kH_cH_m$$
$$k_3 = -nH_c$$
$$k = ml/N$$
l = core length in the direction of magnetization
N = coil turns per phase

And the eddy-current loss density is expressed

$$U = \frac{1}{2} \, rH_n^2 \tag{2.80}$$

To obtain the fundamental of E and δ from Eq. (2.76) is indeed too complicated; instead, a plot for the wave was secured first for a particular reactor having the following data (see Figs. 2.33 and 2.34):

$$l = 1.84 \times 10^{-2} \text{ m}$$
$$a = 2.42 \times 10^{-4} \text{ m}^2$$
$$\mu \approx 2.5 \, \mu$$
$$H_c = 6 \times 10^4 \text{ A-T/m}$$
$$H_m = 9.85 \times 10^4 \text{ A-T/m}$$
$$B_0 = 0.0625 \, \Sigma_0^n \, \frac{1 - \rho^{n+1}}{1 - \rho} \qquad \rho = 0.7$$

The plot gave a nearly odd rectangular periodic waveform, from which the amplitude of the fundamental is calculated. The amplitude of the E

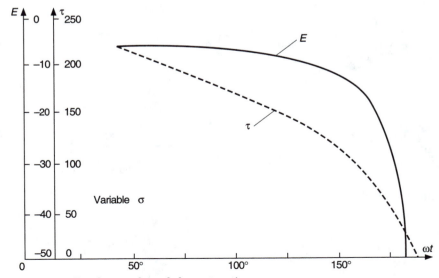

Figure 2.33 E and τ waveforms below saturation.

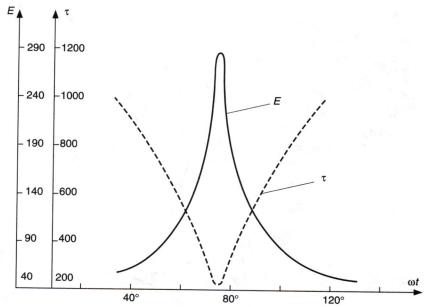

Figure 2.34 E and τ waveforms at saturation.

field = 46 V, while its fundamental peak = 22 V. Then, with respect to the same reactor whose data was mentioned earlier, and taking a constant value of $\sigma = 0.076\ \Omega - m$ (corresponds to the maximum resistive component of the reactor and a peak value of impressed $H_m = 9.85 \times 10^4$ A-T/m and $H_c = 6 \times 10^4$ A-T/m), the eddy-current loss density was calculated to be 126×10^4 w/m^2 and 0.5122 Ω/m surface resistance.

Similarly, for the case where σ is a function of the applied mmf, the eddy-current loss density was calculated to be 93×10^4 w/m^2 and a surface resistance of 0.347 Ω/m. In either case, the level of the magnetic flux density was taken at 0.188 tesla.

Experimental work.   Tests were performed on the same reactor, applying a series of AC voltages and measuring the wattmeter as well as current readings, and after required corrections, the total iron loss/m^3 and the net resistance per phase are obtained and plotted with respect to line current in Fig. 2.35.

Conclusion

1. In a reactor with hard ferromagnetic core characterized with an actual magnetization curve where the B field follows a geometrical series and the H-displaced field follows an arithmetical series, closed-

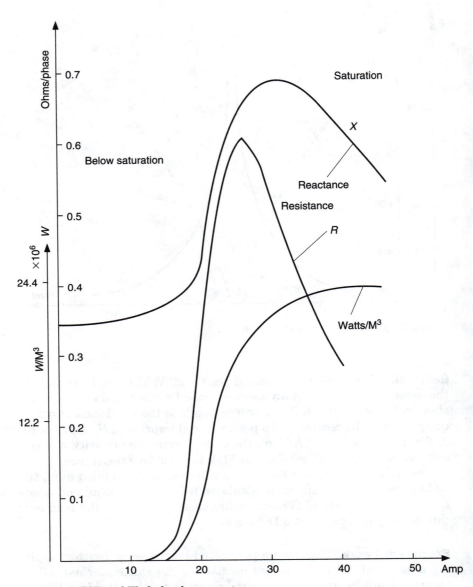

Figure 2.35 R, X, and W of a hard-core reactor.

form solutions have been obtained for the induced electric field, the separating plane, the surface impedance, and the eddy-current loss density. Those solutions represent two cases, the first when σ is constant, and the other for σ as function of the applied mmf.

2. The experimentally measured value of the eddy-current loss density is about 9 percent less than the calculated value with σ as function of the applied mmf, and even less in the case of σ as constant.

3. While this section considered the actual feature of the B-H curve, the calculations were confined only to the case of $B = 0.188$ tesla, corresponding to the application that could be followed at any other value of B, since the field buildup is a series of delayed-step functions.

2.8 Three-Dimensional Model of Eddy-Current Theory in Hard Ferromagnet

In previous work by this author, special theory for eddy-current loss calculation in a hard magnetic cored reactor was established in a single-dimensional mode. Solutions for the induced electric field, propagation of the plane of magnetization, the surface impedance, and the poynting vector were obtained.

Solutions in three-dimensional coordinate systems are sought for the following electromagnetic field components in a hard ferromagnetic continuum known as ALNICO, with the B-H curve a displaced symmetrical step function:

- The induced electric field
- The propagating magnetization plane
- The surface impedance
- Eddy-current loss density

2.8.1 Solutions of the induced field and the propagating plane

Electromagnetic field polarizations in three dimensions are assumed below:

$$e_{xi}, h_y, r_z$$

$$e_x, h_z, r_y$$

$$e_y, h_z, r_x$$

$$e_y, h_x, r_z$$

$$e_z, h_x, r_y$$

$$e_z, h_y, r_x \tag{2.81}$$

where e_x, e_y, and e_z represent the respective applied or exciting magnetic fields which are sinusoidal, as expressed following.

δ_x, δ_y, and δ_z represent the propagating magnetizing phases.

$$h_x = H_{xm} \sin \omega t$$

$$h_y = H_{ym} \sin \omega t$$

$$h_z = H_{zm} \sin \omega t \tag{2.82}$$

By the application of Maxwell's field equation, shown in previous work by this author:

$$\frac{\partial h_x}{\partial Z} = \frac{H_x - H_{cx}}{r_z} = \sigma e_x = \frac{H_y - H_{cy}}{r_z} \tag{2.83}$$

$$\frac{\partial h_y}{\partial x} = \frac{H_y - H_{cy}}{r_x} = \sigma e_y = \frac{H_z - H_{cz}}{r_x} \tag{2.84}$$

$$\frac{\partial h_z}{\partial y} = \frac{H_x - H_{cx}}{r_y} = \sigma e_z = \frac{H_z - H_{cz}}{r_y} \tag{2.85}$$

where H_{xc}, H_{yc}, and H_{zc} represent the magnetic field inception of the coercive force imposed by the B-H curve of the hard ferromagnetic shown in Fig. 2.2a

From the preceding relationships, solutions for the propagating magnetizing plane in three dimensions are expressed:

$$r_x = 2v(t - t_c) \left[\sqrt{\frac{H_{ym}}{\sigma \omega B_{oy}}} + \sqrt{\frac{H_{zm}}{\sigma \omega B_{oz}}} \right] [\cos \omega t_c - \cos \omega t + (\omega t_c - \omega t) \sin \omega t_c]^{\frac{1}{2}} \tag{2.86}$$

Similarly,

$$r_y = 2v(t - t_c) \left[\sqrt{\frac{H_{xm}}{\sigma \omega B_{ox}}} + \sqrt{\frac{H_{zm}}{\sigma \omega B_{oz}}} \right] [\cos \omega t_c - \cos \omega t + (\omega t_c - \omega t) \sin \omega t_c]^{\frac{1}{2}} \tag{2.87}$$

where t_c is the time of the start for the magnetization process in each dimension. Then

$$r_z = 2v(t - t_c) \left[\sqrt{\frac{H_{xm}}{\sigma \omega B_{ox}}} + \sqrt{\frac{H_{ym}}{\sigma \omega B_{oy}}} \right] [\cos \omega t_c - \cos \omega t + (\omega t_c - \omega t) \sin \omega t_c]^{\frac{1}{2}}$$

(2.88)

The depth of maximum penetration for the induced electric field in the x, y, and z dimensions could be secured by setting in Eqs. (2.85), (2.86), and (2.87),

$$\omega t = (\pi - \omega t_c)$$

Therefore

$$\delta_x = r_x \mid \omega t = \pi - \omega t_c$$
$$\delta_y = r_y \mid \omega t = \pi - \omega t_c$$
$$\delta_z = r_z \mid \omega t = \pi - \omega t_c$$

(2.89)

The induced electric field components in three dimensions are expressed following:

$$e_x = 2B_{oy} \frac{\partial r_z}{\partial t}$$

$$= 4B_{oy} \left[\sqrt{\frac{H_{xm}}{\sigma \omega B_{ox}}} + \sqrt{\frac{H_{ym}}{\sigma \omega B_{oy}}} \right]$$

$$= \left\{ v_0(t - t_c)[\cos \omega t_c - \cos \omega t + \sin \omega t_c(\omega t_c - \omega t)]^{\frac{1}{2}} + \right.$$

$$\left. \left[v_1(t - t_c) \frac{\omega(\sin \omega t - \sin \omega t_c)}{2[\cos \omega t_c - \cos \omega t + \sin \omega t_c(\omega t_c - \omega t)]} \right] \right\}$$

(2.90)

Similarly,

$$e_y = 2B_{ox} \frac{\partial r_x}{\partial t}$$

$$= 4B_{oz} \left[\sqrt{\frac{H_{ym}}{\sigma \omega B_{oy}}} + \sqrt{\frac{H_{zm}}{\sigma \omega B_{oz}}} \right]$$

$$= \left\{ v_0(t - t_c)[cos\omega t_c - cos\omega t + sin\omega t_c(\omega t_c - \omega t)]^{\frac{1}{2}} + \right.$$

$$\left. \left[v_1(t - t_c) \frac{\omega(sin\omega t - sin\omega t_c)}{[cos\omega t_c - cos\omega t + sin\omega t_c(\omega t_c - \omega t)]^{\frac{1}{2}}} \right] \right\} \quad (2.91)$$

and

$$e_z = 2B_{ox} \frac{\partial r_y}{\partial t}$$

$$= 4B_{ox} \left[\sqrt{\frac{H_{xm}}{\sigma\omega B_{ox}}} + \sqrt{\frac{H_{zm}}{\sigma\omega B_{oz}}} \right]$$

$$= \left\{ v_0(t - t_c)[cos\omega t_c - cos\omega t + sin\omega t_c(\omega t_c - \omega t)]^{\frac{1}{2}} + \right.$$

$$\left. \left[v_1(t - t_c) \frac{\omega(sin\omega t - sin\omega t_c)}{[cos\omega t_c - cos\omega t + sin\omega t_c(\omega t_c - \omega t)]^{\frac{1}{2}}} \right] \right\} \quad (2.92)$$

where $v_o(t - t_c)$ is a singularity delayed impulse function, B_{ox}, B_{oy}, and B_{oz} represent respective magnetic flux density controlled by the B-H curve as shown in Fig. 2.2a, and $v_1(t - t_c)$ is a delayed step function.

2.8.2   Solution of the surface impedance (Z)

Let Z be the complex surface impedance:

$$Z = R + iX \quad (2.93)$$

Z = max. of fundamental of e/max. of fundamental h.

Graphical illustration for e_x, H_y, and r_z in one state of polarization is shown in Fig. 2.26, where induced electric field is found very close for triangular waveform.

Using the concept of triangular waveform for each of e_x, e_y, and e_z, and expanding through Fourier theorem, the fundamental components are expressed as follows:

$$E_{x\delta} = v(t - t_c) \left[\sqrt{\frac{H_{ym}B_{oy}\omega}{\sigma}} + \sqrt{\frac{H_{xm}B_{ox}\omega}{\sigma}} \right]$$

$$= [(k_1 + k_1')cos\omega t + (k_2 + k_2')sin\omega t] \quad (2.94)$$

$$E_{y\delta} = v(t - t_c) \left[\sqrt{\frac{H_{zm}B_{oz}\omega}{\sigma}} + \sqrt{\frac{H_{ym}B_{oy}\omega}{\sigma}} \right]$$

$$= [(k_1 + k'_1)\cos\omega t + (k_2 + k'_2)\sin\omega t] \qquad (2.95)$$

$$E_{z\delta} = v(t - t_c) \left[\sqrt{\frac{H_{xm}B_{ox}\omega}{\sigma}} + \sqrt{\frac{H_{zm}B_{oz}\omega}{\sigma}} \right]$$

$$= [(k_1 + k'_1)\cos\omega t + (k_2 + k'_2)\sin\omega t] \qquad (2.96)$$

where k, k', k_2, and k'_2 have been listed in equations.

$\theta_m = \omega t$ for maximum τ

$\theta_c = \omega t_c$

a, b, A, and B are constants for the linear function representing the induced electric field in triangular waveform.

$$Z_x = E_{y\delta}/H_{zm} = R_x + jX_x$$

$$Z_y = E_{z\delta}/H_{xm} = R_y + jX_y$$

$$Z_z = E_{x\delta}/H_{ym} = R_z + jX_z \qquad (2.97)$$

Substituting for the $\overline{E}$ and $\overline{H}$ fields into Eq. (2.96).

$$Z_x = \frac{\left[\sqrt{\dfrac{H_{ym}B_{oy}\omega}{\sigma}} + \sqrt{\dfrac{H_{xm}B_{ox}\omega}{\sigma}} \right] [(k_1 + k'_1) + j(k_2 + k'_2)]}{-jH_{zm}} \qquad (2.98)$$

Therefore,

$$R_x = \left[\sqrt{\frac{H_{ym}B_{oy}\omega}{\sigma}} + \sqrt{\frac{H_{xm}B_{oz}\omega}{\sigma}} \right] (k_2 + k'_2) \qquad (2.99)$$

$$X_x = \left[\sqrt{\frac{H_{ym}B_{oy}\omega}{\sigma}} + \sqrt{\frac{H_{xm}B_{oz}\omega}{\sigma}} \right] (k_1 + k'_2) \qquad (2.100)$$

Similarly,

$$R_x = \left[\sqrt{\frac{H_{xm}B_{ox}\omega}{\sigma}} + \sqrt{\frac{H_{zm}B_{oz}\omega}{\sigma}} \right] (k_2 + k'_2) \qquad (2.101)$$

$$X_x = \left[\sqrt{\frac{H_{xm}B_{ox}\omega}{\sigma}} + \sqrt{\frac{H_{zm}B_{oz}\omega}{\sigma}} \right] (k_1 + k'_2) \qquad (2.102)$$

$$R_z = \left[\sqrt{\frac{H_{ym}B_{oy}\omega}{\sigma}} + \sqrt{\frac{H_{xm}B_{ox}\omega}{\sigma}} \right](k_2 + k'_2) \qquad (2.103)$$

$$X_x = \left[\sqrt{\frac{H_{ym}B_{oy}\omega}{\sigma}} + \sqrt{\frac{H_{zm}B_{ox}\omega}{\sigma}} \right](k_1 + k'_2) \qquad (2.104)$$

Consequently, R_t, X_t, as the total surface resistance and reactance, are expressed as:

$$R_t = R_x + R_y + R_z \qquad (2.105)$$

and

$$X_t = X_x + X_y + X_z \qquad (2.106)$$

2.8.3 Poynting vectors ($\overline{W}$)

$$\overline{W}_t = \overline{W}_x + \overline{W}_y + \overline{W}_z \qquad (2.107)$$

$$\overline{W}_x = \frac{1}{2}E_{xf}H_{ym} = \frac{1}{2}R_x(|H_{ym}|)^2 \text{ watts/m}^2 \qquad (2.108)$$

$$\overline{W}_y = \frac{1}{2}E_{yf}H_{zm} = \frac{1}{2}R_y(|H_{zm}|)^2 \text{ watts/m}^2 \qquad (2.109)$$

$$\overline{W}_z = \frac{1}{2}E_{zf}H_{zm} = \frac{1}{2}R_z(|H_{xm}|)^2 \text{ W/m}^2 \qquad (2.110)$$

Therefore, substituting the corresponding $\overline{E}$ and $\overline{H}$ fields into Eqs. (2.106) to (2.108):

$$\overline{W}_x = \frac{1}{2}H_{ym}^2 \left[\sqrt{\frac{H_{ym}B_{oy}\omega}{\sigma}} + \sqrt{\frac{H_{xm}B_{ox}\omega}{\sigma}} \right](K_2 + k'_2) \qquad (2.111)$$

$$\overline{W}_y = \frac{1}{2}H_{zm}^2 \left[\sqrt{\frac{H_{xm}B_{ox}\omega}{\sigma}} + \sqrt{\frac{H_{zm}B_{oz}\omega}{\sigma}} \right](K_2 + k'_2) \qquad (2.112)$$

$$\overline{W}_y = \frac{1}{2}H_{zm}^2 \left[\sqrt{\frac{H_{ym}B_{oy}\omega}{\sigma}} + \sqrt{\frac{H_{xm}B_{ox}\omega}{\sigma}} \right](K_2 + k'_2) \qquad (2.113)$$

Therefore,

$$|W_t| = \{W_x|^2 + |W_y|^2 + |W_z|^2\}^{\frac{1}{2}} \qquad (2.114)$$

W_t represents the average eddy-current loss density in watts/m^2.

This has to indicate that the electrical conductivity (σ) for hard magnetic material is a function of the cored coil line current or, in effect, the external magnetic field intensity. Variation of σ as a function of the field $H_{external}$ covers basically the regime before and after saturation. Hence, the status of σ in the preceding solutions is referred to as an average value σ, or in terms of the resistivity ρ and its average $<\rho>$, where

$$<\rho> = \frac{1}{i_s} \left[\int_i^{i_m} \rho_1(i) + \int_{i_m}^{i_s} \rho_2(i) \right] \qquad (2.115)$$

where i_1 is the initial pickup
   i_m is the current at which ρ is max.
   i_s is the current at saturation

From the preceding presentation, the following conclusions can be drawn:

1. For hard ferromagnetic core material subjected to three-dimensional external magnetic field, the following solutions have been accomplished successfully in three dimensions:

    The advancing magnetization plane

    The induced electric field

    The maximum depth of penetration

    The surface impedances

    The eddy-current poynting vectors

2. Exact computation of total eddy-current loss in any thick, hard magnetic core could be secured by multiplying each dimensional density by the respective surface area.

2.9 Problems

2.1 In reference to Fig. 2.6, derive a mathematical equation for the transient behavior of $i_{ac}(t)$. The applied $V_{ac} = 57.5$ V repeats the process regarding Fig. 2.7, where $V_{ac} = 51.5$ V. Correlate the effect of ΔV_{ac} on the differential $\Delta i_{ac}(t)$ for the two cases. In both cases the residual DC field has been contributed by 27 A.

2.2 In reference to Fig. 2.8, derive a mathematical equation for the transient behavior of $i_{ac}(t)$ where the residual DC field is due to 14 A and the applied $V_{ac} = 57.5$ V. Repeat this process regarding Fig. 2.9, where the residual DC field is due to 40 A and $V_{ac} = 34$ V. Correlate $\Delta i_{ac}(t)$ with respect to $\Delta V_{ac}(t)$ and $\Delta \equiv dc$.

2.3 In reference to Figs. 2.13, 2.14, and 2.15, where the applied $V_{ac} = 6.7$ V and the residual DC fields have been contributed by IDC = 14, 28, and 42 A, establish mathematical representation for $\Delta i_{ac}(t)$ with respect to i_{DC}.

2.4 Repeat the process outlined in Prob. 2.3 for the transient responses of i_{ac} in a saturistor shown in Figs. 2.16, 2.17, and 2.18, where V_{ac} applied is 12.7 V.

2.5 From the transient responses shown in Figs. 2.16, 2.17, and 2.18, give a mathematical equation for the saturistor time-constant with respect to i_{ac} with a plot.

2.6 In reference to the transient responses shown in Figs. 2.19, 2.20, and 2.21, derive a mathematical equation for $\Delta i_{ac}(t)$ with respect to residual DC current. The applied $V_{ac} = 30$ V rms.

2.7 From the result obtained in Prob. 2.6, establish a correlation for the time-constant with respect to the residual DC field.

2.8 In reference to the saturistor transient responses shown in Figs. 2.22, 2.23, and 2.24, establish a mathematical relationship for Δi_{ac} with respect to ΔI_{DC}. The applied $V_{ac} = 43$ V rms.

2.9 From the data secured in Prob. 2.8, obtain mathematical representation for the saturistor time-constant with respect to I_{DC}.

2.10 In reference to Fig. 2.25, plot a curve for ΔR with respect to i_{ac}; then try to obtain a mathematical relationship for such a curve.

2.11 From the data established in Prob. 2.10, plot a curve for $+\Delta R)i_{ac}^2$ with respect to i_{ac}. Indicate condition for maximum power dissipation.

2.12 In reference to Fig. 2.26, plot a curve for ΔX with respect to i_{ac}, and then try to represent the curve by a mathematical relationship.

2.13 In reference to Fig. 2.27, plot a curve for ΔZ with respect i_{ac}, and then try to represent the curve by a mathematical relationship.

2.14 In reference to Fig. 2.28, plot a curve for the differential power density with respect to resistance Δ component for the connected saturistor maximum. Use the data and results from Prob. 2.11.

2.15 In reference to Fig. 2.29, showing resistance plots for the Y-connected saturistor with respect to applied i_{ac}, plot ΔR for cold-hot versus i_{ac}, and then try to obtain an equation representing it.

2.16 In reference to Fig. 2.30, plot ΔX for hot-cold saturistor with respect to i_{ac}, and then try to obtain an equation to represent the graph.

2.17 In reference to Fig. 2.31, plot ΔZ for hot-cold saturistor with respect to i_{ac}, and then try to obtain an equation representing the curve.

2.18 In reference to Fig. 2.32, plot the differential power density with respect to differential saturistor resistance. You may use the data and results of Prob. 2.15.

2.19 Consider a sinusoidal waveform incident at the surface of a thick saturistor core. If the displaced H-H curve for the core is as shown in Fig. 2.36 obtain solutions for the induced electric field, the magnetizing separating plane, and the maximum depth of penetration.

2.20 From the results obtained in Prob. 2.19, establish a solution for the poynting vector surface impedance and the fundamental component of the induced electric field.

2.21 Repeat Prob. 2.19 for the waveform of the core B-H curve shown in Fig. 2.37.

2.22 From the results obtained in Problem 2.21, obtain a solution for the poynting vector and surface impedance.

2.23 For the data given in Sec. 2.6, derive solutions for the induced electric field, and the rate of movement for the magnetizing separating plane inside the saturable-resistor core, if the core is thin such that $d < \delta$, where δ is the maximum expected depth of penetration in a thick core.

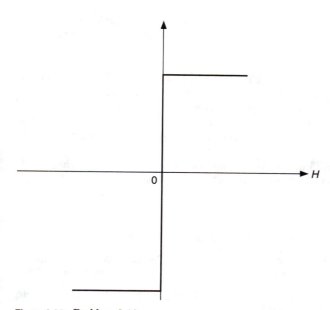

Figure 2.36 Problem 2.19.

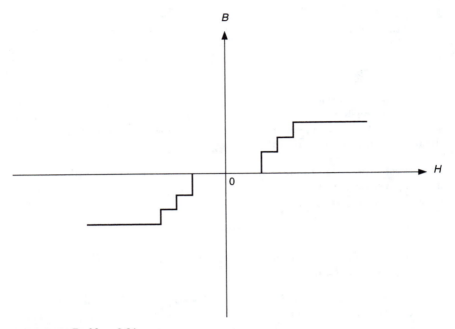

Figure 2.37 Problem 2.21.

2.24 From the results obtained in Prob. 2.23, obtain solutions for the poynting vector and surface impedance, as well as the fundamental component of the induced electric field.

2.25 Solutions are found in Sec. 2.6 for the induced electric field, the magnetizing separating plane, the poynting vector, the surface impedance, and the eddy-current power density for the case where the core electrical conductivity divided by σ is constant. Rederive those solutions if σ is a function of the applied magnetic field $H(t)$, i.e., $\sigma = 1 \backslash n + mH(t)$ as indicated in Sec. 2.6.5, App. B.

2.26 Resolve Prob. 2.25 if the saturable-resistor core is thin such that $d < \delta$, where δ is the maximum depth of penetration in a thick core.

2.10 References

1. Alger, P. L., "A Capacitor Motor with ALNICO Trans Bars in the Rotor, *IEE Power and Apparatus and Systems,* October 1964.
2. Alger, P. L., Coelho, W. A., and Mulund, R. P., "Speed Control of Wound Roto Motors with SCR's and Saturistors," *EEE Trans. on Industry and General Applications,* 1968.
3. Alger, P. L., *Induction Machines,* Gordon and Breach Inc., 1971.
4. Agarwal, P. D., "Eddy Current Losses in Solid Laminated Iron," *AIEE Transactions,* pt. I ("Communications and Electronics"), vol. 78, 1959.
5. Beland, B. "Eddy Currents in Circular, Square and Rectangular Rods," *IEE Proceedings—A,* vol. 130, part A, no. 3, 1983.

6. Balchin, M. J., and Davidson, J.A.M., "3-Dimensional Eddy-Current Calculation by the Network Method Formulation Using Magnetic Scalor potential for non-conducting regions, "*IEE Proceedings,* vol. 130, part A, no. 2, 1983.
7. Balanis, C. A., *Advanced Engineering Electromagnetics,* John Wiley & Sons, 1989.
8. Dekker, A. J., *Solid-State Physics,* Prentice-Hall, 1963.
9. Denno, K., "Eddy-Current Theory in Hard, Thick Ferromagnetic Materials," IEEE—Power Engineering Society Conference Paper C75-005-4, presented at the winter meeting in New York, 1975.
10. Denno, K. I., "Current Limiting in High Voltage Transmission System," *Proceedings of the Canadian Communications and EHV Conference,* 1972.
11. Gunn, C. E., "Improved Starting Performance of Wound-Rotor Motors Using Saturistors," *IEEE Trans. Power Apparatus and System,* 1964.
12. Lynch, A. C., Drake, A.-E.,, and Dix, C. H., "Measurement of Eddy-Current Conductivity," *IEEE Proceedings,* vol. 130, part A, no. 5, 1983.
13. Ramo, S., Whinnery, J., and Duzer, T. V., *Fields and Waves in Communication Electronics,* John Wiley and Sons Inc., 1967.
14. Rodger, D., "Finite Element Method for Calculating Power Frequency—Three-Dimensional Electromagnetic Field Distributions," *IEE Proceedings A,* vol. 130, part A, no. 5, 1983.

Saturable-Resistor in Induction and Synchronous Machines

3.1 Generalized Properties

In this chapter, the role of the saturable-resistor in operational performance and protection of AC machines is to be discussed analytically and, in some cases, experimentally. Effects of this protective electromagnetic device in improving the performance will be examined for the three-phase induction motor of the wound rotor type, the single-phase induction motor, the synchronous motor, the induction three-phase motor of the deep-bar rotor, as well as the conventional linear induction accelerator. In the area of protection, the saturable-resistor is used in the three-phase induction motor of regular wound rotor type, the deep-bar rotor type, and the single-phase induction motor in a design mode intended to limit drastically the starting current, providing a smooth and stable torque over a considerable range of speed, and optimizing the operating ratio of output torque with respect to motor line current.

Reported problems during operation of induction motors pointed to a situation of cutoff of power supply, followed in a few milliseconds by resumption of power to the motor, causing the slip-rings bearing to break down with serious damage to the rotor structural coils. Explanation of this is that, upon interruption of applied voltage followed by resumption of power, if it happens at an angle close to 180°, the total voltage at the slip-rings could in the limit be twice the normal operating voltage, rendering a high inrush current with severe stress at the slip-rings bearing and effective adverse heating subjected at the rotor windings.

The saturable-resistors impedance, reactance, and resistance start to pick up after a small value of AC current along an almost linear path, peaking at the limit of magnetic saturation, after which those ohmic values start to decline downward. Therefore, upon the surge of current at the slip-rings and at the rotor coils, the connected saturable-resistor across the slip-rings offers an automatic increase in ohmic levels that will limit the surge of current in the rotor. Then, upon normalization of operating voltage and with subsequent reduction of the saturable-resistor to the prepickup values, a contacter will automatically disconnect the saturistor from the rotor circuit.

Another solution to contain the damage of surge current on the rotor of induction machines is the connection of a saturable-resistor with DC control at the slip-rings, whereby the DC control is on during normal operation, and a contacter will suspend the control, after which the saturable-resistor will act to limit the flow of current surge produced by the cutoff and the almost immediate return of power supply at the motor terminals.

A similar situation may arise for the AC synchronizing generator throughout the period in which the alternator is passing across the starting cycle as an induction machine or, similarly, in cases where the total voltage is almost twice the normal level, resulting in the addition of the local generated emg and the normally supplied value.

Another protective and control role for the saturable-resistor may be seen as a switching device in the control network of the linear induction accelerator. A LINAC with saturistor can provide charging as well as accelerating beam current of time duration in the order of μsec, which is desired in the utilization of heavy ion beam for inertial fusion research.

Feasibility of using the saturable-resistor in another AC machine (namely, the induction generator) looks promising, where research work carried out by this author indicated widening the scope of operating power-factor from leading to lagging range, steadiness of output power, and significant reduction in the generator internal losses.

In the following presentations, discussion with analysis will address the roles of the saturable-resistor in the function of protection and performance improvements of induction, synchronous machines as well as in linear induction accelerators.

3.2   Response of Rotor Current Due to Slip Change

3.2.1 $I_r(t)$ without saturable-resistor

In this section solution for rotor current $I_r(t)$ in response to a change in slip is to be presented using the rotor circuit subjected to the entire stator voltage as a vehicle to establish the sought solution.

Figure 3.1 shows the rotor circuit at any slip and without saturistor:

$$I_r = \frac{V_s}{jX_r + \dfrac{R_r}{S}} \qquad (3.1)$$

$$|I_r| = \frac{V_s}{\left(X_r^2 + \dfrac{R_r}{S}\right)^{\frac{1}{2}}} \qquad (3.2)$$

Performing the process:

$$\frac{\partial |I_r|}{\partial S} = \frac{\partial |I_r|}{\partial t} = \frac{\partial t}{\partial S}$$

where

$$\frac{\partial |I_r|}{\partial t} = \frac{V_s S'}{[R_r^2 + S^2 X_r^2]^{\frac{1}{2}}} - \frac{S^2 X_r^2 V_s S'}{[R_r^2 + S^2 X_r^2]^{\frac{3}{2}}} \qquad (3.3)$$

On the basis that $S^2 X_r^2 \ll R_r^2$ or in effect $S^4 X_r^4 \ll R_r^4$, the right-hand side of Eq. (3.3) becomes, after application of the binomial expansion theorem:

$$\frac{\partial |I_r|}{\partial t} \approx V_s \left[\frac{S'}{R_r} - \frac{X_r^2}{2R_r^2} S^2 S' - \frac{X_r^2}{R_r^3} S^2 S' + \frac{3}{2} \frac{X_r^4}{R_r^5} S^4 S'' + \cdots \right] \qquad (3.4)$$

where

$$S' = \frac{\partial S}{\partial t} \qquad (3.5)$$

Response of $|I_r|(t)$ produced by sudden changes in the slip(s) is expressed as:

$$1. \quad S(t) = A U_{-1}(t) \qquad (3.6)$$

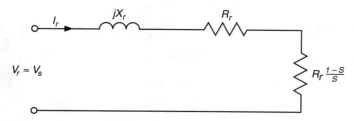

Figure 3.1   Rotor circuit without saturistor.

where $U_{-1}(t)$ is a step function

$$2. \quad S(t) = AU_o(t) \tag{3.7}$$

where $U_o(t)$ is an impulse function, sometimes identified as $\delta(t)$

$$3. \quad S(t) = AU_{-1}(t) - AU_{-1}(t - T) \tag{}$$
$$= A \text{ rectangular pulse} \tag{3.8}$$

Taking the Laplace transform of Eq. (3.4) and securing the inverse transform, the following solutions have been obtained for $|I_r|(t)$.

For $S(t) = AU_{-1}(t)$

$$|I_r|(t) = DU_{-1}(t) + I_o \tag{3.9}$$

where

$$D = \left[\frac{1}{R_r} - \frac{A^2 X_r^2}{2R_r^2} - \frac{X_x^2}{R_r^3} A^2 + \frac{3}{2} \frac{X_r^4}{R_r^5} A^4 \right] \tag{3.10}$$

For $S(t) = AU_0(t)$

$$|I_r|(t) = V_s\beta_1 U_o(t) + V_s\beta_2 U_0'(t) - V_s\beta_3 U_{-1}(t) + I_o U_{-1}(t) \tag{3.11}$$

where

$$\beta_1 = \left[\frac{A}{R_r} - \frac{X_r^2}{2R_r^2} A^2 S_o + \frac{X_r^2 A^2}{R_r^3} S_o - \frac{3}{2} \frac{X_r^4}{R_r^5} A^4 S_o \right] \tag{3.12}$$

where S_o is the initial slip at $t = 0^-$

$$\beta_2 = \left[\frac{3X_r^4}{2R_r^5} A^5 - \frac{X_r^2 A^3}{2R_r^2} \right] \tag{3.13}$$

and

$$\beta_3 = -\frac{S_o}{R_r} \tag{3.14}$$

For $S(t) = AU_{-1}(t) - AU_{-1}(t - T)$

$$|I_r|(t) = (V_s m_1 + I_o)U_{-1}(t) + \tag{}$$
$$V_s m_2 U_{-1}(t - T) + \tag{}$$
$$V_s m_3 U_{-1}(t) \tag{3.15}$$

where

$$m_1 \cong \left[\frac{A - S_o}{R_r} - \gamma_1 (A^3 - A^2 S_o) + \gamma_2 (A^5 - A^4 S_o) \right] \qquad (3.16)$$

$$\gamma_1 = \left[\frac{X_r^2}{2R_r^2} + \frac{X_r^2}{R_r^3} \right] \qquad (3.17)$$

$$\gamma_2 = \frac{3}{2} \frac{X_r^4}{R_r^5} \qquad (3.18)$$

$$m_2 = \left[\gamma_2 (A^4 S_o - 2A^5) - \frac{A}{R_r} - \gamma_1 (A^2 S_o - 2A^3) \right] \qquad (3.19)$$

$$m_3 = [\gamma_1 A^3 + \gamma_2 A^5] \qquad (3.20)$$

3.2.2 $I_r(t)$ with saturable-resistor in the rotor

The saturable-resistor, resistance which changes linearly with frequency, consequently, when it is reflected into the stator it becomes (R/S) $S = R$. S is the slip, while the reactance component of the saturable-resistor (X) is added directly to the leakage reactance of the rotor coils.

Therefore, the rotor region with the addition of the saturistor becomes as shown in Fig. 3.2.

Now we can write:

$$|I_r| = \frac{V_s}{\left[X_r'^2 + \left(R + \frac{R_r}{S} \right)^2 \right]^{\frac{1}{2}}} \qquad (3.21)$$

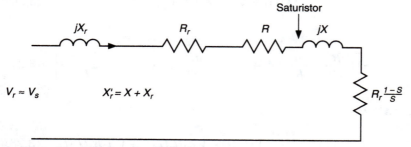

Figure 3.2 Rotor circuit with saturistor.

where

$$X'_r = X_r + X \tag{3.22}$$

and then:

$$\frac{\partial |I_r|}{\partial t} = \frac{V_s R_r R S S'}{[X_r'^2 S^2 + (SR + R_r)^2]^{\frac{1}{2}}} + \frac{V_s R_r^2 S'}{[X_r'^2 S^2 + (SR + R_r)^2]^{\frac{3}{2}}} \tag{3.23}$$

since

$$X_r'^2 S^2 \ll [SR + R_r]^2$$

Using the process of binomial expansion on the two terms in Eq. (3.23), it now becomes:

$$\frac{\partial |I_r|}{\partial t} \approx V_s R_r \left[\left(\frac{R}{R_r^3} - \frac{a}{R_r^5} \right) SS' - \frac{aR}{R_r^6} S^2 S' + \frac{S'}{R_r^2} + \cdots \right] \tag{3.24}$$

where

$$a = 3RR_r^2 \tag{3.25}$$

Response of $|I_r(t)|$ produced by sudden changes in the slip (s) expressed as:

1. $S(t) = AU_{-1}(t)$

2. $S(t) = AU_o(t)$

3. $S(t) = AU_{-1}(t) - AU_{-1}(t - T)$

Taking the Laplace transform with respect to (t) of Eq. (3.24), followed by the process of inverse transform, the following solutions have been obtained for $|I_r(t)|$.

$S(t) = AU_{-1}(t)$

$$|I_r(t)| = V_s R_r (M + I_o) U_{-1}(t) \tag{3.26}$$

where

$$M = \left[\left(\frac{R}{R_r^5} - \frac{a}{R_r^5} \right) A(A - S_o) - \frac{aR}{R_r^6} A^2 (A - S_o) + \frac{A - S_o}{R_r^2} \right] \tag{3.27}$$

$S(t) = AU_o(t)$

$$|I_r(t)| = \left(I_o - \frac{U_x S_o}{R_r} \right) U_{-1}(t) - V_s a_1 R_r U_o(t) + V_s a_2 R_r U'_o(t) \tag{3.28}$$

where

$$a_1 = \left[\frac{A}{R_r^2} - AK_1 S_o + K_2 A^2 S_o \right] \tag{3.29}$$

$$a_2 = [K_1 A^2 - K_2 A^3] \tag{3.30}$$

$$K_1 = \frac{a}{R_r^3} - \frac{a}{R_r^5} \tag{3.31}$$

$$K_2 = \frac{aR}{R_r^6} \tag{3.32}$$

The value of a is shown in Eq. (3.25).

$S(t) = AU_{-1}(t) - AU_{-1}(t - T)$

$$|I_r(t)| \approx V_s R_r(\alpha_1 + \alpha_4) U_{-1}(t) - V_s R_r \alpha_2 U_{-1}(t - T)$$
$$+ V_s R_r \alpha_3 U_{-1}(t - 2T) \tag{3.33}$$

where

$$\alpha_1 = (b_1 A^2 - b_1 AS_o + b_2 A^3 - b_2 A^2 S_o) \tag{3.34}$$

$$\alpha_2 = (2b_1 A^2 - b_1 AS_o + 2b_2 A^3 + \frac{A}{R_r^3} - ab_2 A^2 S_o) \tag{3.35}$$

$$b_1 = \left(\frac{R}{R_r^3} - \frac{\alpha}{R_r^5} \right) \tag{3.36}$$

$$b_2 = \frac{aR}{R_r^6} \tag{3.37}$$

3.3 Induced Torque Response Due to Change in Slip

3.3.1 $T(t)$ without saturistor

$T(t)$ is the induced electromagnetic torque across the air gap and into the rotor.

$T_{ind}(t)$ without saturistor in the rotor

$$T_{ind} = 3 |I_r|^2 \frac{R_r}{S} \tag{3.38}$$

$$I_r = \frac{V_s}{jXr + R_r/S} \tag{3.39}$$

Therefore, from Eqs. (3.38) and (3.39)

$$T_{ind} = 3R_rV_s^2 \frac{1}{S^2X_r^2} + \frac{R_r^2}{S} \tag{3.40}$$

Proceeding with the process,

$$\frac{\partial T}{\partial S} = \frac{\partial T}{\partial t} = \frac{\partial t}{\partial S}$$

Therefore,

$$\frac{\partial T_{ind}}{\partial t} = \frac{3R_rV_s^2 S'}{R_z^2 + S^2X_r^2} - \frac{6R_rV_s^2X_r^2S^2S'}{(R_r^2 + S^2X_r^2)^2} \tag{3.41}$$

Since $S^2X_r^2 \ll R_r^2$, and using the binomial expansion on Eq. (3.41), it becomes

$$\frac{\partial T_{ind}}{\partial t} = 3R_rV_s^2\left[\frac{S'}{R_r^2} - \frac{X_r^2}{R_r^4}S^2S' + \frac{X_r^4}{R_r^4}S^4S'\right] -$$

$$6R_rX_r^2V_s^2\left[\frac{S^2S'}{R_r^4} - \frac{2X_r^2}{R_r^6}S^4S' - \frac{X_r^4}{R_r^4}S^6S'\right]$$

$$+ \cdots \tag{3.42}$$

Solution for $T_{ind}(t)$ for the following sudden changes in the slip (s):

$$S(t) = AU_{-1}(t)$$

$$S(t) = AU_o(t)$$

$$S(t) = AU_{-1}(t) - AU_{-1}(t - T)$$

$S(t) = AU_{-1}(t)$:

$$T_{ind}(t) = 3\left[MU_{-1}(t) + T(0^+)\right] \tag{3.43}$$

where

$$M = \frac{V_s^2}{R_r}(A - S_o)$$

$$-\left(\frac{V_s^2X_r^2}{R_r^3} + 2\frac{V_s^2X_r^2}{R_r^3}\right)A^2(A - S_o)$$

$$+\left(\frac{V_s^2 X_r^2}{R_r^3} + 2\,\frac{V_s^2 X_r^4}{R_r^5}\right) A^4 (A - S_o)$$

$$+2\,\frac{V_s^2 X_r^6}{R_r^3}\,A^6 (A - S_o)$$

$$+\cdots \tag{3.44}$$

$S(t) = AU_o(t)$: Now,

$$T_{ind}(t) = -3h_0 U_{-1}(t) + 3h_1 U_0(t)$$
$$+3h_2 U_0'(t) + T_0 \tag{3.45}$$

where

$$h_1 = [K_0 A + K_1 A^2 S_0 - K_2 A^4 S_0 - K_3 A^6 S_0] \tag{3.46}$$

$$h_2 = [K_1 A^3 + K_2 A^5 + K_3 A^7] \tag{3.47}$$

$$h_0 = K_0 S_0 \tag{3.48}$$

$$K_0 = \frac{V_s^2}{R_r} \tag{3.49}$$

$$K_1 = \left(\frac{V_s^2 X_r^2}{R_r^3} + \frac{2V_s^2 X_r^2}{R_r^3}\right) \tag{3.50}$$

$$K_2 = \left(\frac{V_s^2 X_r^4}{R_r^3} + \frac{2V_s^2 X_r^4}{R_r^5}\right) \tag{3.51}$$

$$K_3 = \frac{2V_s^2 X_r^6}{R_r^3} \tag{3.52}$$

$S(t) = AU_{-1}(t) - AU_{-1}(t - T)$: Solution for $T_{ind}(t)$:

$$T_{ind}(t) = 3H_1 U_{-1}(t) + 3H_2 U_{-1}(t - T) - H_3 U_{-1}(t - 2T) + T_0 \tag{3.53}$$

where

$$H_1 = K_0(A - S_0) - K_1 A^3 + K_1 S_0 A^2 + K_2 A^5$$
$$-K_2 S_0 A^4 + K_3 A^7 - K_3 S_0 A^6] \tag{3.54}$$

$$H_2 = [-K_0 + K_1 A^3 - K_1 S_0 A^2 + K_1 A^3 - K_2 A^5$$
$$-kK_2 S_0 A^4 - K_2 A^5 - K_3 A^7 + K_3 S_0 A^6 - K_3 A^7] \tag{3.55}$$

and

$$H_3 = [-K_1A^3 + K_2A^5 + K_3A^7] \tag{3.56}$$

3.3.2 $T_{ind}(t)$ with saturistor in the rotor

Saturistor impedance in the rotor and reflected into the stator is expressed by:

$$Z = S\,\frac{R}{S} + jX = (R + jX)/\text{phase}$$

$$|I_r^2| = \frac{V_s^2}{(X_r + x)^2 + \left(R + \dfrac{R_r}{S}\right)^2} \tag{3.57}$$

Let $X_r + X = X'$ and since

$$T_{ind} = 3\left[\frac{S^2 V_s^2}{S^2 X'^2 + (R_r + SR)^2}\right]\left[R + \frac{R_r}{S}\right]S' \tag{3.58}$$

Then, carrying out the process of:

$$\frac{\partial T}{\partial S} = \frac{\partial T}{\partial t} = \frac{\partial S}{\partial t}$$

on Eq. (3.58). The $\theta T_{ind}/\theta t$ is given as:

$$\frac{\partial T_{ind}}{\partial t} = \left[\frac{AS^2 + BS + C}{DS^2 ES + F} - \frac{aS^3 + bS^2 + cS}{dS^4 + eS^3 + fS^2 + gS + h}\right]S' \tag{3.59}$$

where $A = -R$
$$B = (2R - R_r)$$
$$C = R_2$$
$$D = (X'^2 + R_r^2)$$
$$E = 2RR_r$$
$$F = R_r^2 \tag{3.60}$$

Also, $a = (2R^3 + 2X'^2 R)$
$$b = 2R_r X'^2 + 4R_r R^2$$
$$c = -2RR_r^2$$
$$d = X'^4 + R^4$$
$$e = 4R_r R^3 + 2RX'^2$$
$$f = 6R_r^2 R^2 + 2R_r X'^2$$
$$g = 4RR_r^3$$
$$h = R_r^4 \tag{3.61}$$

Considering the fact that

$$S^2 X'^2 \ll (R_r + SR)^2$$

and applying the binomial expansion theorem, Eq. (3.59) expressing T_{ind} becomes:

$$\frac{\partial T_{ind}}{\partial t}(s) \approx 3V_s^2 \left[M_1 S^2 + M_2 S - M_3\right] S' \qquad (3.62)$$

where

$$M_1 = \frac{cg/h - b}{h}$$

$$M_2 = \left[\frac{B - CE/F}{F} - \frac{c}{h}\right]$$

$$M_3 = \frac{C}{F} \qquad (3.63)$$

As we did earlier, now consider a sudden change in slip (s) according to the following:

$$S(t) = AU_{-1}(t)$$

$$S(t) = AU_0(t)$$

$$S(t) = AU_{-1}(t) - AU_{-1}(t - T)$$

$S(t) = AU_{-1}(t)$: Therefore,

$$T'_{ind}(s) = 3V_s^2 \left[M_1 A^3 + (M_2 - M_1 S_0)A^2 - M_2 S_0 + M_3 A - M_3 S_0\right] \qquad (3.64)$$

and, therefore,

$$T_{ind}(t) = 3V_s^2[Y]U_{-1}(t) + T_0 \qquad (3.65)$$

where Y represents the terms inside the parentheses of Eq. (3.64).

$S(t) = AU_0(t)$: Solution for $T_{ind}(t)$ from Eq. (3.62) has been obtained as shown:

$$T_{ind}(t) = P_1 U_0'(t) + P_2 U_0(t) + P_3 U_{-1}(t) + T_0 \qquad (3.66)$$

where $P_1 = M_1 A^3 + M_2 A^2$
$P_2 = M_3 A - M_2 A S_0 - M_1 A^2 S_0$
$P_3 = -M_3 S_0 \qquad (3.67)$

$S(t) = AU_{-1}(t) - AU_{-1}(t - T)$:   Referring to Eq. (3.62), by substituting the transform of every term, followed by process of the inverse transform, solution for $T_{ind}(t)$ is given as follows:

$$T_{ind}(t) = 3V_s^2 [g_1 U_{-1}(t) + g_2 U_{-1}(t - T) + g_3 U_{-1}(t - 2T)] + T_0 \quad (3.68)$$

where $g_1 = [M_1(A^3 - A^2 S_0) + M_2(A^2 - AS_0) + M_3 A - M_3 S_0]$
$g_2 = [M_1(A^2 S_0 - 2A^2) + M_2(AS_0 - 2A^2) - M_3 A]$
$g_3 = [M_1 A^3 + M_2]$ $\qquad\qquad\qquad\qquad\qquad\qquad$ (3.69)

3.4   Ratio of Torque/Rotor Current

The most important design and performance parameter for the three-phase induction motor is the optimal ratio of torque with respect to stator line current I_s. However, the rotor current I_r which is slightly less than I_s, could be used in this ratio as reliable replacement. In the previous sections, the author presented sequential response for the induced electromagnetic torque and the rotor current due to sudden increase in motor slip s.

In the following, ratios of torque to rotor current, defined as the ratio Q, for each change in $S(t)$ are given:

3.4.1   Without saturistor in the rotor

For $S(t) = AU_{-1}(t)$

$$Q(t) = \frac{MU_{-1}(t) + T_0}{DU_{-1}(t) + I_0} \qquad (3.70)$$

For $S(t) = AU_0(t)$

$$Q(t) = \frac{-h_0 U_{-1}(t) + h_1 U_0(t) + H_2 U_0'(t) + T_0}{-V_s \beta_3 U_{-1}(t) + V_s \beta_1 U_0(t) + V_s \beta_2 U_0'(t) + I_0} \qquad (3.71)$$

For $S(t) = AU_{-1}(t) - AU_{-1}(t - T)$

$$Q(t) = \frac{H_0 U_{-1}(t) + h_2 U_{-1}(t - T) - H_3 U_{-1}'(t - 2T) + T_0}{V_s m_1 U_{-1}(t) + V_s m_2 U_{-1}(t - T) + V_s m_3 U_{-1}(t - 2T) + I_0} \qquad (3.72)$$

3.4.2   With saturistor in the rotor

For $S(t) = AU_{-1}(t)$

$$Q(t) = \frac{V_s^2 Y U_{-1}(t) + T_0}{V_s R_r (M + I_0) U_{-1}(t)} \qquad (3.73)$$

For $S(t) = AU_0(t)$

$$Q(t) = \frac{P_1 U_0'(t) + P_2 U_0(t) + P_3 U_{-1}(t) + T_0}{\left(I_0 - \dfrac{V_s S_0}{R_r}\right) U_{-1}(t) + V_s a_1 R_r U_0(t) + V_s R_r a_2 U_0'(t)} \tag{3.74}$$

For $S(t) = AU_{-1}(t) - AU_{-1}(t - T)$

$$Q(t) = \frac{V_s^2 [g_1 U_{-1}(t) + g_2 U_{-1}(t - T) + g_3 U_{-1}(t - 2T) + T_0]}{[V_s R_r(\alpha_1 + \alpha_4) + I_0]U_{-1}(t) - V_s R_r \alpha_2 U_{-1}(t - T) + V_s R_r \alpha_3 U_{-1}(t - 2T)} \tag{3.75}$$

3.5   Slip for Maximum Torque with Saturistor

Let the saturistor impedance Z be expressed by

$$Z = R + jX$$

and rotor current I_r by

$$|Ir| = \frac{V_s}{\sqrt{X'^2 + \left(R + \dfrac{R_r}{S}\right)^2}} \tag{3.76}$$

where

$$X' = X_r + X$$

therefore,

$$T_{ind} = 3\,|I_r|^2\,\frac{R_r}{S}$$

$$= 3\,\frac{S V_s^2 R_r}{S^2 X'^2 + (RS + R_r)^2} \tag{3.77}$$

Now carrying out the process of:

$$\frac{\partial |I_r|}{\partial S} = \frac{\partial |I_r|}{\partial t} = \frac{\partial t}{\partial S} = 0$$

From Eq. (3.77), obtaining S_m for maximum T_{ind}:

$$S_m = \frac{RR_r \pm \sqrt{5R^2 R_r^2 + 4R_r^2 X'^2}}{2(R^2 + X'^2)} \tag{3.78}$$

From Fig. 3.3 (repeated here for the convenience of the reader), we may express R and X of the saturable-resistor as rising linear function in terms of rotor current I_r up to the point of saturation, after which they start to decline linearly.

Therefore,

$$R = ai + b = a_1'' + b_1''I_r$$

$$X = a'i + b' = a_2'' + b_2''I_r \qquad (3.79)$$

Substituting the functions for R and X from Eq. (3.79) into Eq. (3.78), differenting S_m with respect to I_r, and letting $b_1'' = b_2'' = 0$, implying pickup value for $I_r = 0$, and considering a, $a' \gg 1$ (slope for R and X are relatively large), the conditional value for I_r at maximum slip for maximum torque is approximately expressed by I_{r-Sm}, given below:

$$I_{r-Sm} \approx \left[\frac{1}{4R_r(5a_1''^2 + 4a_2''^2)} \pm \frac{2}{\sqrt{5a_1''^3 + 4a_2''^2}} \pm \frac{\sqrt{5a_1''^2 + 4a_2''^2}}{a} \right] \qquad (3.80)$$

3.5.1 T_{ind} when saturistor is function of rotor current

In previous sections of this chapter, response of T_{ind} with respect to a sudden change in motor slip (which included a step forcing function,

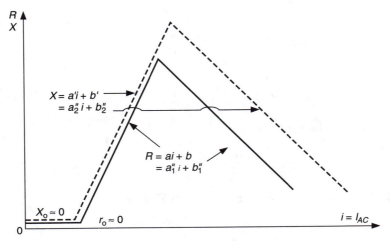

Figure 3.3 Functional representation for R and X.

an impulse, and a pulse) has been calculated with the saturistor impedance considered constant at any slip. In Sec. 3.5, solutions for the motor slip at maximum torque as well as for the rotor current for peak S_m have been secured. Function of the rotor impedance as a linear function of rotor current was assumed.

In this section, general response for T_{ind} with respect to singular, sudden changes in motor slip where the saturistor resistance is a linear function of rotor current will be presented with some acceptable and justifiable approximations.

$$T_{ind} = 3\,|I_r|^2\,\frac{R_r}{S}$$

and

$$I_r = \frac{V_s}{jX_r + \dfrac{R_r}{S} + a'' + j_r b'' + jX} \tag{3.81}$$

where R of saturistor $= a'' + I_r b''$ such that
$R = a_1'' + b_1'' I_r$ and
$X =$ assumed independent of rotor current for the sake of mathematical simplicity and accepted physical reality $\tag{3.82}$

$$|I_r| = \frac{V_s}{\left[\left(\dfrac{R_r}{S} + a'' - b'' I_r\right)^2 + (X_r + X)^2\right]^{\frac{1}{2}}} \tag{3.83}$$

$$T_{ind} = 3\,\frac{V_s^2 S^2}{(R_r - a''S + b'' I_r S)^2 + S^2 X'^2} \tag{3.84}$$

where $X' = X_r + X$

Differentiating T_{ind} in Eq. (3.84) with respect to s, and letting $a'' = 0$, implying the resistance pickup currently to be zero, the following is obtained:

$$\frac{1}{R_r V_s^2}\frac{\partial T_{ind}}{\partial S} = \frac{1}{S^2(X'^2 + b''I_r^2) + S2R_r b'' I_r + R_r^2} -$$

$$\frac{S^2(2X'^2 + b''I_r'^2) + SR_r b'' I_r}{[S^2(X'^2 + b''I_r^2) + S2R_r b'' I_r + R_r^2]^2} -$$

$$\cdots \tag{3.85}$$

Applying a process of long division and neglecting higher-order terms, Eq. (3.85) becomes

$$\frac{\partial T_{ind}}{\partial S} = \left(\frac{\partial T_{ind}}{\partial t}\right)\left(\frac{\partial t}{\partial S}\right) = 0$$

Hence,

$$\frac{\partial T_{ind}}{\partial t} \cong R_r V_s^2 \left[\left(\frac{1}{R_r^2} - S\frac{P_2}{R_r^4}\right)S' + \left(S\frac{P_4}{R_r^4} - S^2\frac{2P_2P_4}{R_r^6}\right)S'\right]$$

or

$$\frac{\partial T_{ind}}{\partial t} \approx R_r V_s^2 \left[\frac{S'}{R_r^2} + \left(\frac{P_4}{R_r^4} - \frac{P_2}{R_r^4}\right)S\,S' - 2\frac{P_2P_4}{R_r^4}S^2S'\right] \quad (3.86)$$

where $P_1 = X_r^2 + b''^2 I_r^2$
$P_2 = 2R_r b I_r$
$P_3 = 2X_r^2 + b''^2 I_r^2$
$P_4 = R_r b'' I_r$ (3.87)

Under singular sudden changes by the slip s, the following solutions for T_{ind} have been developed:

$S(t) = AU_{-1}(t)$

$$T_{ind} = (QR_r V_s^2 + T_0)U_{-1}(t) \quad (3.88)$$

where

$$Q = \left[\frac{1}{R_r^2}(A - S_0) + \left(\frac{P_4}{R_r^4} - \frac{P_2}{R_r^4}\right)(A^2 - AS_0)\right.$$

$$\left. - \frac{2P_2P_4}{R_r^6}(A^3 - A^2S_0)\right] \quad (3.89)$$

$S(t) = AU_0(t)$

$$T_{ind} = [(T_0 - M_0R_rV_s^2)\,U_{-1}(t) + m_1'R_rV_s^2U_0(t) + m_2'R_rV_s^2U_0'(t)] \quad (3.90)$$

This author would like to present the process of $T_{ind}(t)$ when $S(t) = AU_{-1}(t) - AU_{-1}(t - T)$ as a problem for the reader to solve.

3.5.2 Effect of power supply interruption and immediate resumption on the rotor

During the well-known power blackout which occurred in the early 1960s in New York city, followed by power resumption from standstill by generators, heavy-horsepower-rating induction motors failed to restart again, and with this disappointing situation the motor ball bearing broke down. Similar situations have also been reported in other industrial plants, indicating accidental breakdown from failure of induction motors to restart after the resumption of power at the motor terminals. What happened was that after power failure at the motor terminals and the immediate resupply a few milliseconds later, the new induced voltage was adding its own approximate value to the existing voltage that was induced at the rotor region just before power resupply. Consequently, the excessive voltage exerted tremendous damaging stress at the rotor bearings, leading to their destruction. This destructive result can be countered by connecting a saturable-resistor in series with the rotor circuit. Since the rotor resistance and reactance rise sharply with the rise of rotor current, the insertion of the saturistor will lead to a situation where the total rotor impedance will increase automatically upon a surge of total voltage subjected at the rotor in a situation of power supply and resupply.

In the following, a sample case is analyzed where electric power is interrupted at $wt = \pi/2$, and resupply takes place at $wt = 3\pi/2$.

Consider the rotor circuit shown in Fig. 3.1:

Let
$$r = R + \frac{R_r}{S}$$
$$l = L + L_r \tag{3.91}$$

where R and L are the saturistor-resistance and self-inductance, respectively. Now,

$$v_r(t) = i_r(t)r + l\,\frac{di_r}{dt} \tag{3.92}$$

Solving for $i_r(t)$ by first taking the Laplace transform of Eq. (3.92), followed by the inverse transform:

$$i_r(t) = I_{ro}e^{-\frac{rt}{l}} + \frac{\omega V_{r-\max}}{l} \times$$

$$\left[\frac{1}{\omega^2 + \dfrac{r^2}{p^2}}\,\theta^{-\frac{rt}{l}} + \frac{1}{\omega\sqrt{\omega^2 + \dfrac{r^2}{p^2}}}\,sin(wt - \phi) \right] \tag{3.93}$$

where

$$\phi = tan^{-1}\frac{\omega l}{r}$$

The steady-state component in Eq. (3.93) is:

$$I_r \text{ in the steady state} = \frac{V_{r-max}}{\sqrt{\omega^2 l^2 + r^2}} \qquad (3.94)$$

whereby, without saturistor in the rotor,

$$I_{r_1} \text{ (steady state)} = \frac{V_{r-max}}{\sqrt{\omega^2 L_r^2 + \left(\dfrac{R_r}{S}\right)^2}} \qquad (3.95)$$

Then, with saturistor in the rotor,

$$I_{r_2} \text{ (steady-state)} = \frac{V_{r-max}}{\sqrt{\omega^2 (L + L_r)^2 + \left(\dfrac{R_r}{S} + R\right)^2}} \qquad (3.96)$$

In Eq. (3.95) $\omega^2 L_r^2 \ll (R_r ls)^2$ and in Eq. (3.96) $L_r \ll L$. Therefore,

$$I_{r_1} \approx \frac{V_{r-max}}{\dfrac{R_r}{S}}$$

and

$$I_{r_2} \approx \frac{V_{r-max}}{\sqrt{\omega^2 L^2 + \left(\dfrac{R_r}{S} + R\right)^2}}$$

Therefore,

$$\frac{I_{r_2}}{I_{r_1}} \approx \frac{\sqrt{\omega^2 L^2 + \left(\dfrac{R_r}{S} + R\right)^2}}{\dfrac{R_r}{S}} \qquad (3.97)$$

The ratio of I_{r_2}/I_{r_1} is considerably less than unity and, consequently, that is the case in the process of accidental power supply and resupply, respectively. This ratio will also be much less than that ratio in a situation where there is no saturable-resistor in the rotor.

3.6 Mathematical Modeling of Deep-Bar and High-Impedance Rotor Induction Motors with Hard Magnetic Core

This section presents conceptual design of a deep-bar induction motor featured with the insertion of a hard magnetic shell at the top of the rotor slot for the optimization of torque with respect to line current, as well as significant reduction in the stator line current.

Polyphase induction motors of high-impedance rotor and deep-bar rotor with soft ferromagnetic cores are characterized by providing increase in resistance and leakage reactance at low drive in order to reduce the line current and maximize electromagnetic, as well as mechanical, torques. However, severe limitations arise in the way of reducing the line current below the limit of 500 percent without lowering the breakdown torque below the ultimate level of 200 percent.

This section presents new models for the high-impedance rotor and the deep-bar rotor with the effective utilization of hard magnetic cores in the rotors of both machines. The hard magnetic core is made up of an alloy with varying percentages of aluminum, nickel, and cobalt known as ALNICO. Such hard ferromagnetic material possesses the properties of large coercive force, small magnetic permeability, and almost a constant power-factor.

Also, as indicated earlier, a reactor with a hard magnetic core has its resistance and reactance as a direct function of magnetic saturation level, frequency, and the time-varying line current, while a superimposed stationary current (DC) can reduce the final level of saturation.

The main objectives of designing the squirrel-cage rotor with a hard ferromagnet is to increase the total impedance at low drive as well as at starting, leading to drastic reduction of the line current below the 5 per unit (p.u.) value, maintaining a flat response of the electromagnetic torque and, in effect, optimizing the ratio of output mechanical torque with respect to line current.

The objective here is that, given the conceptual design of high-impedance and deep-bar rotor induction motor, the optimization of the ratio of torque to line current at low drive, as well as limiting the inrush current under transient conditions, is required as stipulated:

1. Present a conceptual design model based on the insertion of hard magnetic material at the top of the deep-bar slots.

2. Develop new criterion for the increase in total rotor resistance and reactance under all drives in order to optimize the ratio of torque to line current.

3.6.1 Control of saturable resistance and reactance

With the placement of a hard magnetic shell at the top of the idle steel-and-copper bars in the slot of deep-bar induction motor, the regions below and beyond saturation will be discussed.

The region below saturation: The level of saturation referred to here is that with respect to the saturable-resistor that is part the deep-bar structure.

For the region below saturation, the AC resistance of the saturable-resistor portion is expressed:

$$R = a_1 i_2 + b_1 \tag{3.98}$$

where R = total AC resistance of the saturable-resistor below saturation.

$i_2 = I_{max} \sin \omega t$ as the rotor AC induced current

Let

$$a_1 = \frac{r_{max} - r_{dc}}{i_{zm} - i_p} \tag{3.99}$$

$$b_1 = \frac{r_{max}(i_{zm} - i_p) - i_{zm}(r_{max} - r_{dc})}{i_{zm} - i_p} \tag{3.100}$$

where R_{max} is the maximum AC resistor
r_{dc} is the DC resistance
i_p is the pickup current for magnetization
i_{zm} is the AC current for maximum impedance

The region beyond saturation

$$R_r = a_2 i_2 + b_2 \tag{3.101}$$

where $i_2 = I_2 \sin \omega t$ as the rotor-induced current beyond saturation
R_r = total AC resistance of the saturable-resistor beyond saturation

Also, let

$$a_2 = \frac{R_{max} - r_{sat}}{i_{zm} - i_{sat}} \tag{3.102}$$

$$b_2 = \frac{R_{max}(i_{zm} - i_{sat}) - i_{zm}(R_{max} - r_{sat})}{i_{zm} - i_{sat}} \tag{3.103}$$

where r_{sat} is the resistance at saturation
i_{sat} is the DC induced current in the rotor at saturation

Similarly, the saturable reactance below and beyond saturation could be expressed as:

$$r_s = a_1' i_s + b_1' \qquad \text{below saturation} \tag{3.104}$$

$$\chi_R = a_2' i_2 + b_2' \qquad \text{beyond saturation} \tag{3.105}$$

where

$$a_1' = \frac{R_{max} - \chi_o}{i_{zm} - i_p} \tag{3.106}$$

$$b_1' = \frac{R_{max}\,(i_{zm} - i_p) - i_{zm}\,(R_{max} - r_o)}{i_{zm} - i_p} \tag{3.107}$$

$$a_2' = \frac{\chi_{max} - \chi_{sat}}{i_{zm} - i_{sat}} \tag{3.108}$$

$$b_2' = \frac{R_{max}\,(i_{zm} - i_{sat}) - i_{zm}\,(R_{max} - r_{sat})}{i_{zm} - i_{sat}} \tag{3.109}$$

where r_o is the initial reactance of the saturable-resistor before pickup.

3.6.2 Control of total rotor resistance and reactance

Below saturation and high drive with the insertion of hard magnetic shell at the top of deep-bar slot

Let $R_{ac1} \cong R_{dc}G_1(\alpha d) + [U]R$ \hfill (3.110)

$$\chi_{ac1} \cong jR_{dc}G_2\,(\alpha d) + [U]\chi_s \tag{3.111}$$

where $[U]$ is the associative function in terms of DC flux superimposed on the saturable-resistor core
R_{dc} is the DC resistance of the deep bar alone
d is the total depth of the deep bar
w is the width of the deep bar

$$\alpha \text{ is proportional to } \sqrt{\frac{f}{\rho}} \tag{3.112}$$

f is the rotor induced frequency
ρ is the rotor deep-bar resistivity

Below saturation and low drive

$$R_{ac2} \cong R_{dc}G_3(\alpha d) + [U]R \tag{3.113}$$

$$\chi_{ac2} \cong jR_{dc}G_3(\alpha d) + [U]\chi_s \tag{3.114}$$

Beyond saturation and high drive

$$R_{ac3} \cong R_{dc}G_1(\alpha d) + [U]\, R_r \tag{3.115}$$

and

$$\chi_{ac3} \cong jR_{dc}G_2(\alpha d) + [U]\chi_\chi \tag{3.116}$$

Beyond saturation and low drive

$$R_{ac4} \cong R_{dc}G_3(\alpha d) + [U]R_r \tag{3.117}$$

and

$$\chi_{ac4} \cong jR_{dc}G_3(\alpha d) + [U]\chi_R \tag{3.118}$$

In the preceding equations, high drive is constrained by the condition:

$$\alpha d < 1.5 \tag{3.119}$$

while low drive is constrained by the condition:

$$\alpha d > 2 \tag{3.120}$$

Also, the AC resistance of the saturable-resistor below and beyond saturation R_s and R_r, respectively, is a linear function of the rotor-induced field frequency, i.e.,

$$R_s < f$$

and

$$R_r < f$$

3.6.3   Incremental rotor resistance and reactance

High drive

$$\Delta_1 R = KR_{dc}\,[G_1(\alpha d) - 1] + [U]\, R \cdot R_r \tag{3.121}$$

$$\Delta_2 X = KR_{dc}\,[G_2(\alpha d) - 1] + [U]\, \chi_s \cdot \chi_R \tag{3.122}$$

where R_0 is the rotor resistance for the deep bar alone

$$k = \frac{X}{R_0}$$

Low drive

$$\Delta_3 R = K\,[G_3(\alpha d) - 1]\,R_{dc} + [U]\,R \cdot R_r \qquad (3.123)$$

$$\Delta_4 X = K\,[G_3(\alpha d) - 1]\,R_{dc} + [U]\,\chi_s \cdot \chi_R \qquad (3.124)$$

From the preceding equations, the performance ratio for the optimized design and protective operation for the deep-bar rotor with the addition of the thin-shell hard magnetic core could be secured by $\Delta X/\Delta R$ in each of the cases presented.

We note from the preceding discussion that insertion of a thin shell of hard magnetic material at the top of the deep-bar rotor slot of induction rotor will lead to the following practical benefits:

1. Effective control of inrush-induced current in the rotor following rupture in power supply, especially with large machines where high-frequency rotor field is prevailing.

2. Control for the level of total increase in the rotor resistance and reactance at low drive will produce maximum flat torque while at the same time reducing the motor line current and optimizing the ratio of torque to line current.

3. Addition of the thin shell of hard magnetic material at the top of the deep bar will result in relative reduction of the total depth for the bar due to the substantial incremental effect in ohmic levels added by the hard magnetic material. This will result in total reduction in motor size while producing optimization in the machine-torque-to-line-current ratio.

Results presented in this section are of extremely high importance for rendering this new design of the high-impedance and the deep-bar rotor to be more efficient in low-drive applications and in curbing excessive inrush current at starting and during transient behavior due to a sudden short-circuit occurrence.

3.7 Control of Hysteresis Motor through Saturistor in the Rotor

This section presents the analytical domain for the control of speed performance of the single-phase hysteresis motor, which could be secured by the behavior of a saturable-resistor inserted in the rotor. By virtue of the automatic change in the saturable-resistor ohmic values with respect to changes in frequency and current, control of the motor operational performance will ensure optimum outputs for torque and line current.

Control of operational performance of single-hysteresis-phase induction motors of various designs is centered on maximizing the ratio of torque output with respect to the motor main-line current. Present design modes for optimizing the ratio of torque/current output includes high-impedance rotor, deep-bar rotor, idle bar rotor, double squirrel-cage rotor, reduced-voltage starting, hysteresis motor with total ALNICO bars, and other conventional design. All of these preceding designs will contribute positively to one aspect of the motor operations, while generating adverse effect regarding other performance outputs.

Composite rotor design for the single-phase hysteresis motor is based on double-bar structure, with the ALNICO material occupying the top part of the composite rotor bar.

ALNICO material inserted within direct paths of magnetic flux linkage will follow the attitude of the saturable-resistor, whereby the ohmic values of resistance, reactance, and impedance will start to increase sharply after the initiation of the pickup regime, and then proceed to drop at magnetic saturation.

Concurrently, the saturable-resistor power factor and magnetic permeability follow a similar pattern of rise from the level of pickup current up to saturation, and then a decline after that.

3.7.1   The double revolving field theory

In terms of the presence of forward field and backward field circuits in the rotor continuum of the single-phase hysteresis motor, its equivalent circuit referred to the stator for an equal number of turns for the stator and rotor windings is as shown in Figs. 3.4 and 3.5.

In Fig. 3.4

B_f is the terminal-induced emf across the forward dynamic resistance $(R_x + R_r/S)/2 = R_f$.

R_x is the ALNICO-AC resistance.

R_r is the regular DC resistance of squirrel-case rotor.

B_b is the terminal emf across the backward dynamic resistance.

$$\frac{R_x}{2} + \frac{R_r}{(2-S)} = R_b \qquad (3.125)$$

where S is the forward slip,

$$S = \frac{f_s - f_r}{f_s} \qquad (3.126)$$

f_s is the frequency of the stator field in H_z, while f_r corresponds to the rotor mechanical speed.

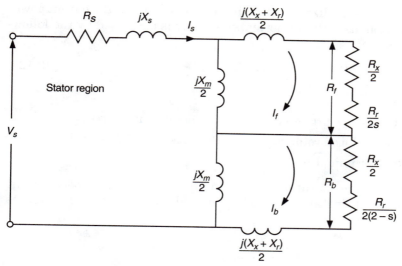

Figure 3.4 Equivalent circuit of single-phase hysteresis motor.

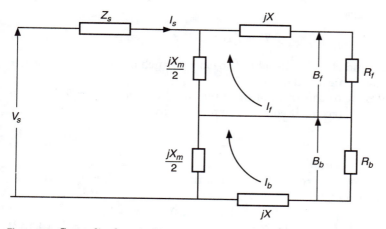

Figure 3.5 Generalized equivalent circuit of single-phase motors.

Also, Eq. (3.126) could be expressed in terms of speeds as shown following:

$$S = \frac{n_s - n_r}{n_s} \qquad (3.127)$$

where

n_s is the speed of the stator rotating field known as the synchronous speed in rpm.

n_r is the rotor speed in rpm.

Now, we can realize from the circuit in Fig. 3.5 that the output power in synchronous watts at normal frequency is expressed by the following equation:

$$T = I_f^2 R_f - I_b^2 R_b \qquad (3.128)$$

where

I_f is the induced rotor current in the forward field circuit

I_b is the induced rotor current in the backward field circuit

s = the per unit slip

However, the pulsation torque due to the interaction between the forward field flux and the backward field current, and vice versa, is given by:

$$T_A = \sum_{n=1}^{N} I_{nj} B_{nb} + \sum_{n=1}^{N} I_{nb} B_{nf} \qquad (3.129)$$

where the index n refers to the order of rotor circuits, including those accounting for impressed sources. And, since the impressed voltage at the rotor is zero,

$$T_A = I_f B_b + I_b B_f \qquad (3.130)$$

where I_f, I_b, and B_b are related to the close branch of the rotor to the air gap, as shown in Figs. 3.4 and 3.5

Turning next to the ratio of unidirectional or average torque to the total machine current, it could be expressed by

$$\frac{T}{I_r} = \frac{I_f^2 R_f - I_b^2 R_b}{I_r} \qquad (3.131)$$

while a similar ratio relating the pulsating torque with respect to total line current is given by:

$$\frac{T_A}{I_r} = \frac{|I_f B_b + I_b B_f|}{I_r} \qquad (3.132)$$

Of course, design control calls for the maximization of T/I_r and minization of T_A/I_r.

Performance control for the single-phase induction hysteresis motor throughout the running condition has the target of optimization for the ratio of the unidirectional torque with respect to total line current. The net output of the unidirectional torque is the result of the difference between the output from the forward-revolving field circuit and that of the backward-revolving field circuit. At the same time, the machine will release the pulsating double-frequency torque output produced by the interaction between rotor current flowing in the forward field circuit with the backward field circuit induced voltage and that produced by the interaction between the rotor backward field current with the forward for induced voltage.

From the preceding presentation, maximization of the unidirectional torque, minimization of the pulsating double-frequency torque, and subsequent relative reduction of the motor line current are the principal elements for the performance optimization of the single-phase induction hysteresis machine. Toward attainment of optimal operation, automatic control is sought for delivering heavy torque with relatively lower line current at low-operation drive and also for higher breakdown torque.

To accomplish the aforementioned improvements, special hysteresis characteristics will be added to the single-phase induction hysteresis machine by designing the rotor of deep-bar mode with ALINCO material to occupy the upper portion of the deep-slot rotor.

Based on the characteristics of ALNICO as special hysteresis material, its role in improving the motor performance is based on the following:

1. Rotor resistance will surge relatively higher in the forward field circuit than in the backward field circuit in such a way that will result in an increase in the unidirectional torque, with subsequent reduction in the line current and the double-frequency pulsating torque. This is because the ALNICO resistance is a function of both frequency and current, as indicated earlier.

2. Since the rotor circuit relevant to ALNICO material will pattern a continuous change in the power factor (hence, magnetic permeability regimes will result in a reduction of leakage reactance that will contribute to significant increase in the breakdown torque), it varies inversely with respect to total rotor leakage reactance.

3.7.2 Summary

We can summarize from the previous discussion the following:

1. Performance improvement for the single-phase hysteresis induction motor could be secured by designing this machine as two modes of hysteresis motor through the addition of ALNICO at the upper portion of a deep, regular hysteresis bar. Operational improvement is basically the optimization of the ratio of the unidirectional torque to the motor line current.

2. Attempt to minimize the adverse effect of the backward field rotor circuit toward relative increase of the unidirectional torque and significantly reduce the double-frequency pulsating torque.

3. Automatic protection for this machine will be provided whenever the supply voltage is disconnected and then restored following cycles afterward, where the total voltage subjected on the rotor will be amplified, with resulting increase in current flow. Under such conditions, ALNICO impedance will rise automatically and, consequently, curb destructive increase in the rotor current.

4. Since ALNICO material characteristics vary in terms of area spread of their hysteresis loops, indicating broad spectrum in coercive force as well as similar reduction in magnetic induction, and since the pattern of the surge in their ohmic values with respect to line current varies considerably, a singular or even multiplicity of several ALNICOS will provide automatic and effective working mode for controlling the optimal performance of electric motors.

Increase of coercive force will lead to delay regarding the initial pickup current and, of course, the commencement of the state of magnetization (and the reverse) will enhance the initiation of magnetization.

Controlling the initiation of magnetization can impose an effective mechanism to provide optimal output torque at any desired speed level, especially at low drive regimes and, at the same time, substantially reduce the total motor line current.

Multiplicity of various modes of ALNICO into a single compound will provide a scope of selectivity for the initial state of magnetization and, hence, a sequential decision for securing desired breakdown torque at any speed level required.

3.8   Feasibility of the AC Induction Generator with Hard Ferromagnetic Core in the Rotor

The presentation in this section advances the prospects of the AC induction generator as an operational power source by integrating a saturistor in the machine rotor.

Calculations for the steady-state aspects will indicate an increase in the range of the generator power factor from leading to lagging levels, steadiness of its power output, and significant reduction in the generator internal power losses.

A dynamic model in the complex frequency domain is developed which identifies the marked effect of the generator time-incremental current change with respect to the input negative slip.

The AC induction generator has the advantage of operating over the whole speed range with no stability problems, no separate DC source needed to excite the rotor, and no system to carry the excitation current to the rotor. Its present disadvantages are (1) that it can deliver output power only in the vicinity of unity-leading power factor, and (2) the need of condenser system to provide the initial exciting current and any external lagging power-factor load.

Improving the prospects of the induction generator as a reliable power source, a closed-core reactor with hard ferromagnetic core is considered to be integrated within the rotor circuit. This reactor is characterized by an impedance which is a direct function of the AC line current, its frequency, and a DC bias (as additional control element).

Refer to Fig. 3.6, which represents the induction generator equivalent circuit with a saturable-resistor inserted in the rotor.

It is necessary to examine the following.

1. Steady-state performance parameters such as: rotor power developed, line current, power factor, and internal dissipation with respect to slip

2. Development of a dynamic model to show the effectiveness of the saturistor on the generator operation

3.8.1   Analytical solutions

Steady state. Performance calculations were carried out on a three-phase induction generator having the following parameters (valid at 50°C):

Stator: $R_1 = 1.10 \ \Omega$, $X_1 = 1.60 \ \Omega$

		Current	R_r	R_h	$X_2 + X_h$
Rotor:	Below pickup	4 A	1.6	0.10	3.70
	Max. impedance	12 A	1.6	1.60	4.90
	Max. current	20 A	1.6	0.80	4.10

$X_m = 40 \ \Omega$

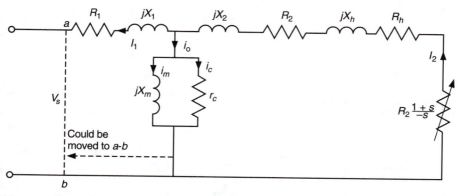

Figure 3.6 Equivalent circut of induction generator.

Steady-state calculation resulted in securing information which is indicated in Figs. 3.7, 3.8, and 3.9. The dependent variable parameter is the generator negative slip(s).

Dynamic model. Using Fig. 3.6 as the base for establishing the required model, the following differential equation is obtained:

$$(R_1 + R_h) + \frac{di/dt}{i} (L_1 + L_2 + L_h - L_m) = \frac{R_2}{s} \tag{3.133}$$

where s is the generator slip.

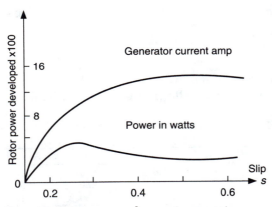

Figure 3.7 Rotor power and generator current.

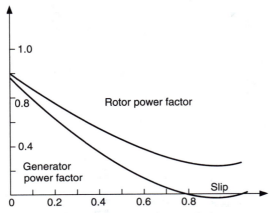

Figure 3.8 Rotor and generator power factors.

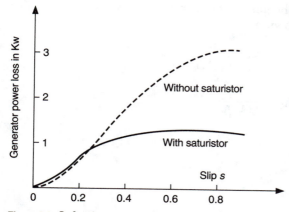

Figure 3.9 Induction generator power loss.

Transformation of Eq. (3.133) results in the following model equation in the complex frequency domain:

$$\frac{F(s)}{H(s)} = -\frac{R_2}{(L_1 + L_2 + L_h - L_m)} - \frac{R_1 - R_h}{(L_1 + L_2 + L_h - L_m)S} \times \frac{1}{H(s)} \quad (3.134)$$

where R_h, L_h = saturistor resistance and inductance, respectively, for the hysteresis motor.

where $H(s) = \text{Transform of} \left(\dfrac{R_2}{s}\right)$

$F(s) = \text{Transform of } \dfrac{di/dt}{i}$, or

$$F(s) = \sum_{k=1}^{Q} \frac{A_1 S_k}{B'_1 S_k} I(S - S_k) \quad (3.135)$$

$\dfrac{A_1(s)}{B_1(s)} = \text{Transform of } di/dt$

$S_k = \text{poles of } B_1(s)$

The dynamic model is shown in Fig. 3.10.
Now we can summarize the following:

1. With hard ferromagnetic core in the generator rotor, AC power output could be delivered over a wide range of power factors, from leading to lagging levels. Also, stable flatness of power output and reduction in internal losses are indicated.

2. Development of the dynamic model indicates the significant effect of the resistive part of the saturable-resistor in stabilizing the time-

incremental factor of the generator current with respect to slip. For Fig. 3.10, note the following:

$$K_1 = \frac{R_2}{L_1 + L_2 + L_h - L_m}$$

$$K_2 = \frac{R_1 + R_h}{L_1 + L_2 + L_h - L_m}$$

$$F(s) = \mathcal{L}\,\frac{di/dt}{i}$$

$$H(s) = \mathcal{L}\,\frac{R_2}{s}$$

3.9 Security of Uninterrupted AC Power through Hard Magnetic Core in the Alternator Damper

Alternators pass through limited state of induction while starting, as well as during abnormal situations that may be forced on the generator during fault incidents. This section presents modification in design procedure intended to secure automatic protection as well as continuity of power supply during the period of time the machine is forced to run as induction generator.

Accidental lines opening to the alternator during induction period may coincide with a situation of phase aiding between the internal and the existing external emfs, resulting in amplification of the total voltage level exerted on the alternator.

Amplification of the induced emf will lead to the injection of high-current inrush into the alternator, especially at low drive, with severe damaging consequences.

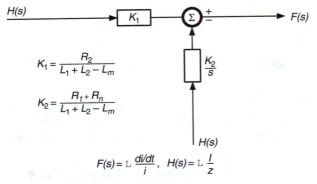

$$F(s) = \mathrm{L}\,\frac{di/dt}{i}, \quad H(s) = \mathrm{L}\,\frac{I}{z}$$

Figure 3.10 Dynamic model of induction generator.

Design modification called for in this section involves the insertion of a thin layer of hard magnetic material at the top of the damper winding in the armature structure.

Magnetic properties of hard materials are characterized by high coercive force, low magnetic permeability, and by an almost unity power factor as a reactor core. Also, it behaves as a hysteresis material where the resistive element increases linearly with frequency. Another very important property is that, as a core of a solenoid unit, its ohmic values are a special function of the line current, where they increase sharply for current values beyond the pickup level, and then start to decline above magnetic saturation. Hence, the presence of the hard magnetic layer at the top of the damper winding will provide the alternator in the induction period with an automatic increase in the armature impedance as a result of accidental line opening or any short-circuiting fault that may force the synchronous generator into the induction state.

An automatic surge in the armature impedance due to the insertion of hard magnetic material layer at the top of the damper winding will curb drastically the rise of the inrush armature current produced by the amplification of terminal emf and the accidental reduction of the generator speed, leading to a surge in the frequency of the induced emf. The automatic increase in the armature impedance will disappear as soon as the normal state of synchronism is returned to the alternator and the state of induction is ceased.

Results identified in this section include modification in the synchronous generator equivalent circuit due to the insertion of hard magnetic material on top of the damper winding, behavior of the generator-induced and terminal emf, and modification in the power-angle equation and the state of voltage regulation modification by the resistive and inductive components introduced by the hard magnet reactive and resistive effects.

Design modification of the alternator structure with hard magnetic material embedded in the damper will ensure automatic uninterruptible supply of AC power, since correct operational electromagnetic state for the alternator will return in a few milliseconds by virtue of the action enforced by the magnetic properties of the hard magnetic material.

3.10   Damping of Inrush Current in Synchronous Generator during State of Induction

This section presents an analytical picture regarding damping action on the synchronous inrush current during the state of induction. Again, this could be accomplished by inserting a hard magnetic core at the top of the damper winding. Modification in the machine-dynamic equations and the equivalent circuit will be presented with attention to the existence of nonzero slip. Steady-state properties of the hard magnetic core act as an automatic switch for releasing the surge in the damper impedance, as well as in stopping it.

Synchronous generators pass through a limited state of induction operation during starting as well as during abnormal situations that may be forced on the machine, especially due to fault conditions.

The intention here is to present analytical design procedure intended to provide automatic protection of a synchronous generator during the period of time the machine is running as an induction rotary device. Accidental lines opening to the synchronous generator during the induction period, followed by reclosing in a short time, may coincide with a situation where the terminal voltage may be additive to the internal phase on induced voltage such that an amplified level of voltage will develop, which could inject high current inrush into the entire generator structure at low drive, resulting in severe damage.

The design-protective procedure called for in this section is the insertion of a thin layer of hard magnetic material at the top of the damper winding in the armature structure.

Properties (as presented earlier) of the hard magnetic material in the magnetic domain are characterized by large coercive force, low magnetic permeability, and by an almost unity power factor as a reactor core. Also, it behaves as a hysteresis material where the resistive element increases linearly with frequency. Another very important property is that as a core of solenoid unit, its ohmic values are a special function of the line current, where they increase sharply for current values beyond the pickup level and then start to decline above magnetic saturation. Hence the presence of a hard magnetic layer at the top of the damper winding will provide the synchronous machine in the induction period with an automatic increase in the armature impedance as a result of accidental line offering or any fault that may force the time-synchronous generator into the induction state.

An automatic surge in the armature impedance due to the insertion of a hard magnetic material layer at the top of the damper winding will curb drastically the rise of the inrush armature current produced by the amplification of terminal emf and the accidental reduction of the generator speed, leading to a surge in the frequency of the induced emf. The automatic increase in the armature impedance will disappear as soon as the normal state of synchronism is returned to the generator and the state of induction is ceased.

Results identified in this section include modification in the synchronous equivalent circuit due to the insertion of hard magnetic material at the top of the damper winding, the generator-induced and terminal emf, and modification of power-angle equations and the state of voltage regulation modification by the resistive and inductive components introduced by the hard magnet.

The synchronous machine during starting, as well as during accidental lines disconnection, will enter into the state of induction through sizable steady-state time duration.

The intention is to establish the following:

1. Modification in the machine steady state and dynamic equations due to the insertion of hard magnetic material at the top of the damper. This will in turn alter the steady-state turn-on of the field and damper impedances in terms of percentage slip (s).

2. New, equivalent circuit reflecting on the existence of the hard magnetic core in the damper during induction.

3.10.1 Modification in the steady-state and dynamic equations of the salient pole machine

Let Z_h be the hard magnetic core impedance:

$$Z_h = r_h + \frac{\partial X_h}{\partial t} \tag{3.136}$$

where $r_h = f(i_D, f_r)$
$X_h = g(i_D, f_r)$
$= 2\pi f_r L_h \tag{3.137}$

i_D is the induced current in the damper.

f_r is the induced current frequency in the rotor damper such that:

$$f_r = s f_s \tag{3.138}$$

S is the slip during induction.

f_r is the stator field frequency.

Refer to Fig. 3.11, which shows the winding sequence of the salient pole synchronous machine with hard magnetic core in the damper.

Next, winding inductances are listed below:

L_{rr} as the rotor self-inductance

$$= L_f + L_D$$

$$= L_r - L_h(f_r \cdot i_D) \tag{3.139}$$

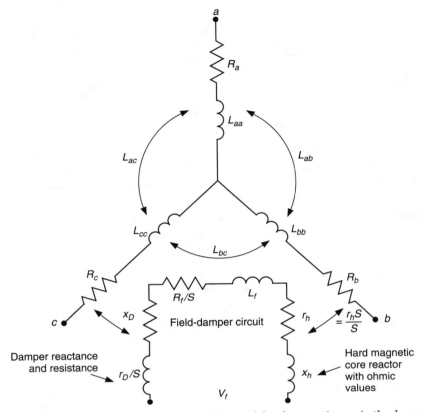

Figure 3.11 Salient pole synchronous machine with hard magnetic core in the damper.

L_r is the self-inductance without hard magnetic core in the damper.
L_h is the hard magnetic core self-inductance.

$$L_{ra} = L_o cos\theta \div L_h(f_r, i_D)cos\theta$$

$$L_{rb} = L_o cos(\theta - 120) + L_h(f_r, i_D)cos(\theta - 120)$$

$$L_{rc} = L_o cos(\theta - 240) + L_h(f_r, i_D)cos(\theta - 240) \tag{3.140}$$

where the angle θ is the rotor magnetic axis with respect to the stator-phase axis due to saliency.

L_{ra}, L_{rb}, and L_{rc} are the mutual inductances between the rotor and the three respective phases.

L_o is the mutual inductance for cylindrical rotor.

Voltage vectors across the stator and rotor could be listed as follows:

$$e_a = -i_a R_a - \frac{\partial}{\partial t}(L_{aa} i_a) - \frac{\partial}{\partial t}(L_{ab} i_b) - \frac{\partial}{\partial t}(L_{ac} i_c) + \frac{\partial}{\partial t}(L_{ra} i_f)$$

$$e_b = -i_b R_b - \frac{\partial}{\partial t}(L_{ba} i_a) - \frac{\partial}{\partial t}(L_{bb} i_b) - \frac{\partial}{\partial t}(L_{bc} i_c) + \frac{\partial}{\partial t}(L_{rb} i_f)$$

$$e_c = -i_c R_c - \frac{\partial}{\partial t}(L_{ca} i_a) - \frac{\partial}{\partial t}(L_{cb} i_b) - \frac{\partial}{\partial t}(L_{cc} i_c) + \frac{\partial}{\partial t}(L_{rc} i_f)$$

$$e_r = i_r \left[\frac{R_f}{S} + \frac{r_D}{S} + r_h(f_r \cdot i_D) \right] - \frac{\partial}{\partial t}(L_{ra} i_a) - \frac{\partial}{\partial t}(L_{rb} i_b)$$

$$- \frac{\partial}{\partial t}(L_{rc} i_c) + \frac{\partial}{\partial t}[L_f + L_h(f_r \cdot i_b)]i_f \tag{3.141}$$

where L_{ra}, L_{rb}, and L_{rc} are expressed in Eq. (3.141). Those inductances reflect on the mutual interaction between the rotor (containing the field excitation circuit, the damper, and the saturable-resistor) and each phase of the stator structure. Implicitly, rotor-stator inductive coefficients account for the presence of damper and hard magnetic core reactor in addition to the field excitation parameters in the rotor circuit.

An AC reactor with hard magnetic core is characterized by the following:

- The resistance, as well as its reactance, are functions of frequency and line current. Dependence on frequency in a linear fashion identifies the resistive component as hysteresis material. Dependency on line current beyond the pickup level will enhance the ohmic values in an almost linear trend up to saturation, after which they will decline again to the prepickup levels.

- It has large A-T/meter coercive force and low magnetic permeability (a few times μ_0).

- It behaves at a circuit condition at an almost unity power factor.

- Magnetic saturation could be controlled by the superposition of DC flux, and hence all the ohmic values will decline in nonlinear order—especially the resistive element from which evolved the name of saturable-resistor, or saturistor. (See Fig. 3.14.)

In the direct axis D

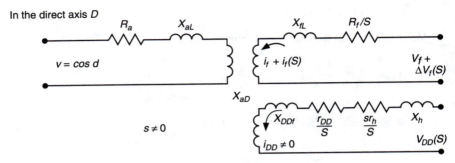

Figure 3.12 Equivalent circuit in the D axis.

In the quadrature axis Q

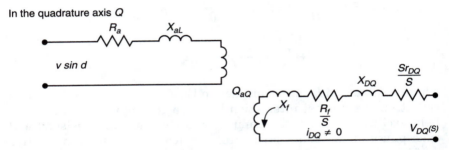

Figure 3.13 Equivalent circuit in the Q axis.

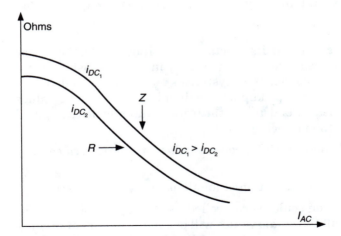

Figure 3.14 Control of ohmic values with DC bias.

The new L matrix of the synchronous machine becomes:

$$L = \begin{bmatrix} L_{aa} & L_{ab} & L_{ac} & L_{ar} \\ L_{bc} & L_{bb} & L_{bc} & L_{br}(f_r, i_D) \\ L_{ac} & L_{cb} & L_{cc} & L_{cr}(f_r, i_D) \\ L_{ra}(f_r, i_D) & L_{rb}(f_r, i_D) & L_{rc}(f_r, i_D) & L_{rr}(f_r, i_D) \end{bmatrix} \quad (3.143a)$$

Similarly, the R matrix is:

$$R = \begin{bmatrix} R_a & 0 & 0 & 0 \\ 0 & R_b & 0 & 0 \\ 0 & 0 & R_c & 0 \\ 0 & 0 & 0 & \left[\dfrac{r_b}{S} + \dfrac{R_f}{S} + R_h(f_r, i_D) \right] \end{bmatrix} \quad (3.143b)$$

3.10.2   Modification in the equivalent circuit

Since the synchronous machine during the induction period will act according to a slip(s) between the stator field frequency and the rotor counterpart, voltage will be induced in the $D + Q$ axis, as shown in the new equivalent circuit in Figs. 3.12 and 3.13.

Now it is time to present the conclusions which follow.

Insertion of hard magnetic core material as a top layer of the damper element in a synchronous machine will lead to the subsequent beneficial performances.

During the period of steady-state induction which may occur as a result of line or lines of disconnection with nonzero slip, or perhaps high slip, the addition of voltage induced internally to the supply value upon reclosing may produce almost duplication of emf level at the machine terminals. The presence of hard magnetic core at the damper will ensure:

1. Total impedance will rise sharply in the damper and hence in the field circuit.

2. Increase in the rotor impedance is due to nonzero high slip and an initial surge in rotor current.

3. The automatic surge in the damper impedance will curb the initial steady-state level of rotor current inrush.

4. Phasor voltage equations as well as rotor emf are modified to account for the effect of slip with nonzero currents in the damper circuits in the D and Q axis.

5. New form for the synchronous machine equivalent circuit is presented, accounting for the state of induction due to the nonzero slip.

6. The state of induction will cease automatically upon return of the machine emf to normal level by virtue of the hard magnet curbing action and the restoration of zero slip state and, of course, synchronism.

7. While the presence of saturable-resistor in the damper is beneficial in curbing the synchronous inrush current during line open circuit, provision has to be designed for limitation of temperature rise generated by the hysteresis property of the saturable-resistor material.

3.11   Synchronized Switching Mechanism for the Linear Induction Accelerators

A feasibility study for the introduction of the saturable-resistor as a switching device in the control network of linear induction accelerators will be discussed. The new accelerator can provide charging as well as accelerating beam current in time duration on the order of μsec, which is very much desired in the utilization of heavy ion beam for inertial fusion research.

Developments in inertia fusion reactor research point to the need for providing beams of heavy ions characterized by relatively long pulse duration, low ion kinetic energy, and several kiloamperes current. This section presents potential design improvement in the spark gap switching mechanism for the linear induction accelerator, including the core type as well as the line type (radial pulse conical, radial pulse cylindrical, and the autocoaxial). The pulse switching involves the utilization of the saturable-resistor with hard magnetic material core. Switching behavior of the saturable-resistor is associated with an open-circuit impedance at synchronized inrush AC exciting current and high frequency, while its ohmic value approaches almost a short circuit just at AC current pickup value and beyond saturation. Calculations have been carried out with explicit results concerning design and performance parameters (spectrum of energy, current, pulse duration, core loss, and surge impedance) as well as synchronizing switching current and frequency for various modes of linear induction accelerators coordinated with an integrated saturable-resistor switching mechanism. Also the saturable-resistor impedance as well as its resistance and reactance increase exactly linearly with frequency. Hence, exciting beam high repetition rate is central for effective utilization of this kind of reactor.

3.11.1   Review of magnetization characteristics of the saturable-resistor

For reader convenience, characteristics of the saturable-resistor are repeated in coordination with the linear induction accelerator as follows.

Surface characteristics impedance Z_s

$$Z_s = \frac{E_{induced\text{-}fundamental}}{H_{applied}}$$

$$Z_s = r + jx \tag{3.144}$$

$$r = -\sqrt{\frac{\omega B_s}{\sigma H_m}} \; (k_2 + k_2')$$

$$x = \sqrt{\frac{\omega B_s}{\sigma H_m}} \; (k_1 + k_1') \tag{3.145}$$

where H_m = amplitude of applied magnetic field; k_1, k_1, and k_2 are constants dependent upon the angular phase for initial magnetization as well as field penetration properties of the saturable-resistor; ω is the exciting beam frequency; B_s is the level of magnetic saturation; and σ is the reactor electrical conductivity expressed by:

$$\sigma = \frac{1}{a i_{a+b}} \tag{3.146}$$

i_c = beam charging current where a and b are constant functions in terms of maximum and minimum ohmic values of the switching saturable-resistor.

Pickup current for magnetization or relaxation

$$i_{rms} = \frac{2.02}{\sqrt{2}} \; \frac{H_c l}{N} \tag{3.147}$$

where H_c = coercive force in ampere-turns/inch
   l = length of magnetic path in inches
   N = number of turns

Levels of H_c for three saturistors with different ALNICO cores are as follows:

ALNICO	5–7	700	oersted min
ALNICO	8	1280	oersted min
ALNICO	9	1500	oersted min

Time rate of decay of transient field. Experimental work conducted on a saturable-resistor with ALNICO 5–7 core, revealed that the entire transient field component for an applied 60-Hz source diminished within 27 ms and its time constant at 13.5 ms. Those time durations were totally independent of premagnetization levels for an applied DC field. Also, it is found that time duration for pulse front rise at 60 Hz is on the order of about 3.75 ms. Therefore, it is expected that the combined effects of frequency and pulse current strength will stretch the powering beam pulse duration to the order of μsec. instead of ns.

3.11.2   Switching mechanism for induction accelerators

From the preceding discussion concerning the complex dependence of all ohmic values for the saturable-resistor with respect to the amplitude of the applied field as well as the linear dependence on frequency, this device can feasibly offer real promise as an operational switch for all models of induction accelerators.

For the core model accelerator, this device acts as a closing switch with almost negligible impedance for powering beam current equal to or less than the saturistor pickup current, while throughout the duration of the exciting pulse, the saturistor impedance increases so quickly that at the peak of the pulse, the saturistor impedance will be extremely high approaching an open-circuit and, hence, terminate the duration of the powering beam pulse.

Also the switching performance of the saturable-resistor could feasibly be used in a similar pattern in the process of magnetic relaxation in the core induction accelerator for resetting the core to the level of residual induction in the demagnetization period.

For the switching performance of line-type induction accelerator, where the principle of induction of the accelerating field is based on changing with time the cross-sectional area of enclosed magnetic flux, the saturable-resistor automatic switching-on-and-off operation is similar to that of core type accelerator, with the added constraints on the changing area by the saturable-resistor surge impedance, which must be compatible with that of the line accelerator (Z_0 of accelerator $= Z_s$ of the saturable-resistor).

3.11.3   Revised forms of induction accelerators: basic equations

The saturable-resistor functions as an automatic switch of the powering beam current, which requires close and careful interconnection with the accelerator network, resulting in the required modifications of the basic performance acceleration equations.

Core-type linear induction accelerator

$$i_c \text{ at saturation} = \frac{\pi d^2}{4\rho_{total}} (\sqrt{ba} + b) \qquad (3.148)$$

$$\rho_{total} = \rho_{core} + \rho_{saturistor} \qquad (3.149)$$

$$\rho_{saturistor} = \frac{A}{l} \frac{E_{induced\ max.}}{H_{applied\ max.}} \qquad (3.150)$$

where i_c = core current
ρ = electrical resistivity
d = magnetic core foil thickness of accelerator
b, a = inner and outer radius of accelerator core

$$V\tau_s(\Delta B)A = \frac{2\sqrt{b}}{\sqrt{a} + \sqrt{b}} \text{ volt} - \text{seconds} \qquad (3.151)$$

where τ_s = saturation time-constant.
$\Delta B = B_s + B_r$ will be controlled by the magnetization and relaxation behavior of the switching pattern of the saturable-resistor.

However, during demagnetization resetting forced by the saturable-resistor hysteresis loop in which B_r is almost equal to B_s, ΔB will approach $2B_s$, resulting in a larger induced field in a similar magnetization process with a change from $-B_r$ to $+B_s$.

Therefore,

$$V\tau_s \approx 2B_sA \frac{2\sqrt{b}}{\sqrt{a} + \sqrt{b}} \text{ volt} - \text{seconds} \qquad (3.152)$$

Line-type linear accelerators. This type includes pulse line accelerator and electron autoaccelerator.

Pulse line accelerator could be either of the radial conical type or the radial cylindrical.

Radial conical accelerator. Its characteristic impedance Z_0 is expressed by:

$$Z_0 = 120ln cot \frac{\theta}{2} \Omega$$

$$\theta = 90° + \frac{\phi}{2}$$

$$\phi = \text{angular inclination of the core} \qquad (3.153)$$

Again, since powering beam switching on and off is feasibly expected to be carried out by the extreme change in the impedance of the saturable-resistor, the new constraint on Z_0 is the perfect matching with the surface impedance Z_s expressed in Eq. (3.144).

Z_s is a function of the pulse line frequency as well as the saturation level of magnetization and the pattern of field penetration through the ALNICO core.

Therefore, from Eqs. (3.144) and (3.153)

$$\theta = 2cot^{-1}d \, \frac{Z_s}{120} \qquad (3.154)$$

According to Eq. (3.154), θ has to be set with respect to the total magnetization behavior of the saturable-resistor switching performance set by the requirement of impedance matching.

Radial cylindrical accelerator. Its characteristic impedance Z_0 is expressed by:

$$Z_0 = 60ln \, \frac{c}{b} = 60ln \, \frac{b}{a} \qquad (3.155)$$

where a, b, and c are coaxial radii of the cylindrical accelerator channels. Similarly, Z_0 must be perfectly matched with Z_s of the saturable-resistor in order to avoid completely any transverse field. Therefore,

$$\frac{c}{b} = \frac{b}{a} = e^{-\frac{4s}{60}} \qquad (3.156)$$

Electron autoaccelerator. Its characteristic impedance Z_0 is expressed by:

$$Z_0 = 60ln \, \frac{a}{b} = \frac{V_a}{I_c - I_b} \qquad (3.157)$$

where V_a = accelerating voltage
I_c, I_b = charging current and beam current, respectively
a, b = outer and inner radius of the coaxial cylinders

Similarly, the constraint imposed by the saturable-resistor switching mechanism is generated by the principle of impedance matching between Z_0 and Z_s. Therefore,

$$\frac{a}{b} = e^{\frac{Z_s}{60}} \qquad (3.158)$$

and

$$V_a = Z_s(I_c - I_b) \qquad (3.159)$$

For maximum conversion efficiency, the beam current

$$I_b = \frac{V_a}{2Z_s} \qquad (3.160)$$

In all types of line accelerators mentioned, the charging current I_c has a duration in the order of μsec., provided by the magnetization and relaxation characteristics of the switching function of the saturable-resistor. Spreading the pulse duration of the charging current is essential for effective inertial fusion research efforts.

To summarize:

1. Switching performance of saturable-resistor for core-type induction accelerator provides well-defined automatic closing and opening switching by virtue of the extreme change in its characteristic impedance.

2. Impedance change of this switching device in all types of induction accelerators provides a reliable stretch of charging pulse time duration of the order of μsec, which is desired in the field of accelerating heavy ions beam in inertial fusion.

3. The principle of impedance matching is central for interconnecting the saturable-resistor in accelerator network.

3.12   Appendix

Throughout Chap. 3, operations for Laplace transforms involved product of functions of changes in induction motor slip (s). Details of such transforms are given following with a fair level of systematic steps.

From the theory for the transform of product of functions indicated

$$\mathcal{L}f_1(t)f_2(t) = \sum_{g=1}^{k} \frac{A_1(S_k)}{B_1'(S_k)} \, F_2(S - S_k) \tag{A.1}$$

where

$$\mathcal{L}f_1(t) = \frac{A_1(S_k)}{B_1(S_k)} \tag{A.2}$$

Equation (A.2) stipulates that the transform of $f_1(t)$ must be a rational function containing a number of first order poles from $g = 1$ to $g = k$.

$$S(t) = AU_{-1}(t)$$

$$\mathcal{L}S(t) = \frac{A}{S} \tag{A.3}$$

$$\mathcal{L}S'(t) = SF(s) - f(0)$$
$$= A - S_0 \tag{A.4}$$

where S_0 is the slip at $t = 0^+$

For LSS': applying the rule given in Eq. (A.1), where

$$\mathcal{L}f_1(t) = F_1(s) = \frac{A}{S}$$

$$\mathcal{L}f_2(t) = F_2(s) = (A - S_0)$$

Therefore,

$$\mathcal{L}S(t)S'(t) = A(A - S_0) \tag{A.5}$$

$$\mathcal{L}S^2(t) = \mathcal{L}A^2U_{-1}(t)$$

$$= \frac{A^2}{S} \tag{A.6}$$

$$\mathcal{L}S^2(t)S'(t) = A^2(A - S_0) \tag{A.7}$$

$$\mathcal{L}S^4(t) = \mathcal{L}A^4U_{-1}(t) = \frac{A^4}{S} \tag{A.8}$$

Therefore,

$$\mathcal{L}S^4(t)S'(t) = A^4(A - S_0) \tag{A.9}$$

Also

$$\mathcal{L}S^n(t)S'(t) = A^n(A - S_0) \tag{A.10}$$

$S(t) = AU_0(t)$, an impulse of magnitude A

$$\mathcal{L}AU(t) = A = A\,\frac{S}{S}$$

$$\mathcal{L}S'(t) = SF(s) - S_0$$

$$= SA - S_0 \tag{A.11}$$

$$\mathcal{L}S^2(t) = \mathcal{L}A^2U_0(t) = A^2 = A^2\,\frac{S}{S}$$

$$\mathcal{L}S^3(t) = \mathcal{L}A^3U_0(t) = A^3 = A^3\,\frac{S}{S}$$

and

$$\mathcal{L}S^n(t) = \mathcal{L}A^nU_0(t) = A^n = A^n\,\frac{S}{S} \tag{A.12}$$

Therefore, by applying the rule of Eq. (A.1):

$$\mathcal{L}S(t)S'(t) = AS(AS - S_0)$$
$$= (A^2S^2 - ASS_0) \tag{A.13}$$

$$\mathcal{L}S^2(t)S'(t) = A^2S(AS - S_0)$$
$$= A^3S^2 - A^2SS_0 \tag{A.14}$$

$$\mathcal{L}S^3(t)S'(t) = A^3S(AS - S_0)$$
$$= A^4S^2 - A^3SS_0 \tag{A.15}$$

and

$$\mathcal{L}S^n(t)S'(t) = A^nS(AS - S_0)$$
$$= A^{n+1}S^2 - A^nSS_0 \tag{A.16}$$

$S(t) = AU_{-1}(t) - AU_{-1}(t - T)$

$$\mathcal{L}S(t) = \frac{A}{S} - \frac{A}{S} e^{-TS} \tag{A.17}$$

$$\mathcal{L}S'(t) = (A - Ae^{-TS}) - S_0 \tag{A.18}$$

$$S^2(t) = A^2U_{-1}(t) - A^2U_{-1}(t - T) \tag{A.19}$$

$$\mathcal{L}S^2(t) = \frac{A^2}{S} - \frac{A^2}{S} e^{-TS} \tag{A.20}$$

$$\mathcal{L}S^2(t)S'(t) = (A^2 - A^2e^{-TS})[A - Ae^{-TS} - S_0] \tag{A.21}$$

$$\mathcal{L}S^3(t)S'(t) = (A^3 - A^3e^{-TS})[A - Ae^{-TS} - S_0] \tag{A.22}$$

and

$$\mathcal{L}S^n(t)S'(t) = (A^n - A^ne^{-TS})[A - Ae^{-TS} - S_0] \tag{A.23}$$

3.13 Problems

3.1 A three-phase wound rotor induction motor with P as the number of poles, 60 Hz running at no-load, when suddenly subjected to a change in slip is characterized by:

$$S(t) = at - b$$

Obtain solutions for the rotor current and for the electromagnetic torque as a function of time.

3.2 From solutions for the rotor current and torque secured in Prob. 3.1, obtain their numerical response after $t = 2$ seconds, and for the following specific data:

$$U_{L-L} = 208 \text{ V}$$
$$p = 8$$
$$R_s = 0.20 \ \Omega$$
$$R_r = 0.12 \ \Omega$$
$$X_m = 15 \ \Omega$$
$$X_s = 0.40 \ \Omega$$
$$X_r = 0.52 \ \Omega$$
$$P_{f+w} = 200 \text{ W}$$
$$P_{core} = 300 \text{ W}$$

3.3 A three saturable-resistor with its resistance and reactance as a function of rotor current I_r (shown in Fig. 3.15) is inserted in the rotor of a three-phase induction motor. The motor is operating at no-load when a sudden change in slip occurred, expressed by:

$$S(t) = at - b$$

Obtain the expression for the rotor current and T_{ind}, considering the surge in R and X of the saturistor is confined to a range in rotor current of $0 < I_r \leq I_{rm1}$ for the resistance and $0 < I_r \leq I_{rm2}$ for the reactance.

3.4 Repeat the process to obtain expressions for $I_r(t)$ and $T_{ind}(t)$ for $I_r > I_{rm}$ for

$$S(t) = at - b$$

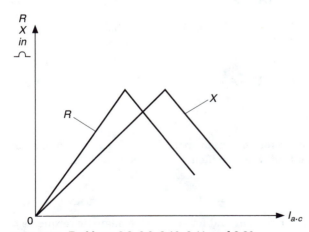

Figure 3.15 Problems 3.3, 3.9, 3.10, 3.11, and 3.20.

3.5 Repeat Prob. 3.1 when the sudden change in slip is characterized by:

$$S(t) = A\sin\omega t$$

3.6 Repeat Prob. 3.1 when the sudden change in slip is characterized by:

$$S(t) = AU_{-1}(t) + A\sin\omega t$$

3.7 Repeat Prob. 3.3 for

$$S(t) = A\sin\omega t$$

3.8 Repeat Prob. 3.4 for

$$S(t) = AU_{-1}(t) + A\sin\omega t$$

3.9 Repeat Prob. 3.3 when R and X of the saturable-resistor changes according to Fig. 3.15, where

$$I_0 = \text{the pickup current}$$

3.10 Repeat Prob. 3.5 when R and X of the saturable-resistor follows the pattern of Fig. 3.15.

3.11 Repeat Prob. 3.6 when R and X of the saturable-resistor follows the pattern of Fig. 3.15.

3.12 A single-phase induction motor is in running condition, having a saturable-resistor inserted in the rotor circuit whose R and X change according to the pattern shown in Fig. (3.3P). Upon a sudden change in slip (s) characterized by

$$S(t) = at - b$$

obtain the expression for the unidirectional torque as a function of time. Consider

$$0 \le I_r \le I_{rm1} \quad \text{for } R$$

and

$$0 \le I_r \le I_{rm2} \quad \text{for } X$$

3.13 Repeat Prob. 3.12, but for the expression of the time response of the double-frequency pulsating torque.

3.14 Repeat Prob. 3.12, but for the time response of the rotor current $I_r(t)$.

3.15 Repeat Prob. 3.12, but for when the range of I_r is beyond I_{rm1} for R, and beyond I_{rm2} for X.

3.16 Repeat Prob. 3.12 when the change in S is given by:

$$S(t) = A \sin \omega t$$

3.17 Repeat Prob. 3.12 when the change in S is given by:

$$S(t) = AU_{-1}(t) + A \sin \omega t$$

3.18 Repeat Prob. 3.13 when the change in S is given by:

$$S(t) = A \sin \omega t$$

3.19 Repeat Prob. 3.13 when the change in S is given by:

$$S(t) = AU_{-1}(t) + A \sin \omega t$$

3.20 Repeat Prob. 3.13 to obtain the time response of the rotor current in the forward field circuit and the backward field circuit. Consider the saturistor resistance and reactance follows the pattern of Fig. 3.15, and

$$0 \le I_r \le I_{rm1} \quad \text{for } R$$

and

$$0 \le I_r \le I_{rm2} \quad \text{for } X$$

3.21 Repeat Prob. 3.20 for the case of $I_r > I_{rm1}$ and $I_r > I_{rm2}$.

3.22 Repeat Prob. 3.20 for the case where:

$$S(t) = A \sin \omega t$$

3.23 Repeat Prob. 3.20 for the case where:

$$S(t) = AU_{-1}(t) + A \sin \omega t$$

3.24 In a deep-bar, three-phase induction, obtain the expression for the induced electromagnetic torque with respect to time when a sudden change in the motor slip (s) is expressed by the motor running at no-load:

$$S(t) = AU_{-1}(t) - AU_{-1}(t - T)$$

Assume that there is no saturable-resistor in the rotor.

3.25 Obtain the expression for the rotor current as a function of time in a three-phase deep-bar induction motor containing a saturable-resistor in the rotor. Consider that the rise of the saturable-resistor impedance is taking place from the pickup level of rotor current up to saturation. The solution of $I_r(t)$ is with respect to a change in slip given by:

$$S(t) = AU_0(t)$$

where initially the motor was running at no-load.

3.26 Repeat Prob. 3.25 where the range of change of the saturable-resistor is confined to the region beyond the peak value.

3.27 Repeat Prob. 3.25 when the sudden change in motor slip is given by:

$$S(t) = A sin\omega t$$

3.28 Repeat Prob. 3.26 when the sudden change in motor slip is given by:

$$S(t) = A cos\omega t$$

3.29 In reference to Prob. 3.25, obtain the expression for the ratio of $I_r(t)$ with no saturable-resistor to that with saturable-resistor in the rotor.

3.30 In a single-phase hysteresis motor, obtain the expression for the forward and backward rotor currents as a function of time with saturable-resistor inserted in the rotor upon a sudden change in slip given by:

$$S(t) = AU_{-1}(t)$$

with the motor initially running at no-load.

3.31 In a single-phase hysteresis motor, obtain the expression for the double-frequency pulsating torque with respect to time in response to a sudden change in slip (as given following) with a saturable-resistor inserted in the rotor circuit. Consider increase in the saturable-resistor and reactance confined to the range from pickup value of rotor current to that at peak ohmic values.

$$S(t) = A sin\omega t$$

3.32 Repeat Prob. 3.31 when the motor rotor current is larger than that corresponding to peak saturistor ohmic values.

3.33 Refer to Eq. (3.143), which shows the inductance matrix of synchronous machines with saturable-resistor in the damper. Approximating the ohmic characteristics of the saturable-resistor shown in Fig. 3.9 as straight-line functions, establish the specific form of the synchronous machine L matrix of Eq. (3.143a).

3.34 Repeat Prob. 3.3 with respect to the R matrix of synchronous machine given by Eq. (3.143b).

3.35 Modify the core current in a core-type linear induction accelerator given by Eq. (3.148) when the contained saturable-resistor: (1) has a displaced B-H curve, and (2) for an actual B-H curve. Comment on any difference.

3.36 Modify the equation for the beam current of the electron autoaccelerator given by Eq. (3.160) to account for Z_s due to a displaced B-H curve, and then for an actual B-H curve. Comment on any difference.

3.14 References

1. R., Alger, *The Nature of Induction Machines,* Gordon and Breach, New York, 1965.
2. P., Alger, G. Angest, and G. Schweder, "Saturistors and Low Starting Current Induction Motors," *Electrical Engineering,* vol. 81, Dec. 1962, pp. 965–969.
3. P. O. Alger, R. L. Mester, and J. G. Yoon, "A Capacitor Motor with Alnico Bars in the Rotor," IEEE Conf. paper presented at the winter meeting of Power Engineering Society, 1963.
4. P. L. Alger, "The Dilema of Single-Phase Induction Motor Theory," conference paper presented at the summer meeting of the AIEE Power Engineering Society, 1958.
5. K. Denno, "Current Limiting in High Voltage Transmission Systems," *Proceedings of the Canadian Communications and EHV Conference,* 1972.
6. K. Denno, "Eddy Current Theory in Hard Thick Ferromagnetic Materials," conference paper no. C75-005-4 presented at the 1975 winter meeting of the IEEE Power Engineering Society.
7. K. Denno, "The Transient Behavior of Saturable Frontiers of Power," Technology conf., Sept. 1971, pp. 18–23.
8. K. Denno, "Special Eddy-Current Theory for Tokamak Fusion Reactor," *Proceedings of the 8th Symposium on Engineering Problems of Fusion-Research,* vol. 1, 1979, pp. 85–88.
9. K. Denno, "Hysteresis Loss Calculations for the Tokamak Fusion Reactor Through Fields Harmonic Content" *Journal—Nuclear Technology/Fusion,* vol. 4, no. 2, part 3, Sept. 1983, pp. 1368–72.
10. K. Denno, "Current-Limiting in High Voltage Transmission System," *Proceedings IEEE Canadian Communication and Power Conference,* 1972.
11. K. Denno, "Calculation of Electro-magnetic Field and Poynting Vector in Thin Hard Ferromagnetic Material," *Proceedings of 4th Reno Conference on Magnetic Fields,* 1973.
12. K. Denno, "Eddy Current Theory in Hard Thick Ferromagnetic Materials," IEEE Conference paper no. C75-005-4 presented at the 1975 winter meeting of the IEEE Power Engineering Society.
13. K. Denno, "Eddy Current Theory for Hard Ferromagnetic Core Reactor with Exact Magnetization Curve." *Proceedings of the 1976 Midwest Power Symposium,* Oct. 1976.
14. K. Denno, "Validity of the Double-Revolving Field and Cross Field Theories for Single Phase Induction Machines with Impedance Comensation," *Proceedings of the 1977 Midwest Power Symposium.*
15. K. Denno, "Modelling of Idle-Bar Rotor Induction Motor with Hard Magnetic Core," *Proceedings of International Coil Winding Association,* 1983, pp. 20–24.
16. K. Denno, "Three Dimension Model of Eddy-Current Theory in Hard ferromagnetic Material," *Proceedings of ISATED International Symposium,* 1984, pp. 13–15.
17. K. Denno, "Mathematical Modelling of Deep-Bar and High Impedance Rotor Induction Motors with Hard Magnetic Core," *Proceedings of ISATED International Symposium,* 1984, pp. 13–15.
18. K. Denno, "Damping of Inrush Current in Synchronous Generator During State of Induction," *Proceedings of IEEE-Industry Applications Society Conf.,* Oct. 1985, pp. 854–58.
19. K. Denno, "Hysteresis Loss Calculations for the Tokamak Fusion Reactor Through Fields Harmonics Content," *Journal-Nuclear Technology/Fusion,* vol. 4, no. 2, part 4, 1983, pp. 1368–72.
20. K. Denno, "Modelling of Idle-Bar Rotor Induction Motor with Hard Magnetic Core," will appear soon in the *International Journal of Energy Systems,* 1985.
21. O. I. Elgerd, *Electric Energy Systems Theory,* 2d edition, McGraw-Hill Book Co., 1982.
22. O. I. Elgerd, *Electric Energy Systems Theory,* McGraw-Hill Book Co., 1971.
23. Eccleshall, D., and Temperley, J. K., "Transfer of Energy from Charged Transmission Lines with Applications to Pulsed High-Current Accelerators," *J. of Applied Physics,* July, 1978.

24. C. E. Gunn, "Improved Starting Performance of Wound-Rotor Motors Using Saturistors," Conf. paper presented at the EEE winter meeting of 1962.

25. V. Gourishankar, D. H. Kelly, *Electro-mechanical Energy Conversion,* 2d edition, Intext Educational Publishers, New York, 1973.

26. G. C. Jain, *Design, Operation and Testing of Synchronous Machines,* Asia Publishing House, 1966.

27. Keefe, D., "Linear Induction Accelerator Conceptual Design," Lawrence Berkeley Laboratory, H-I, FAN-58, 1978.

28. I. L. Kosow, *Electric Machinery and Transformers,* Prentice Hall, Inc., Englewood Cliffs, N.J., 1972.

29. Leiss, J. E., "Induction Linear Accelerators and Their Applications," *IEEE Transactions on Nuclear Science,* vol. NS-26, no. 3, 1979.

30. MIT Staff, *Magnetic Circuits and Transformers,* John Wiley & Sons, Inc., 1943.

31. S. A. Nasar, L. E. Unnewehr, *Electromechanics and Electric Machines,* John Wiley & Sons, Inc., 1979.

32. M. A. Plonus, *Applied Electromagnetics,* McGraw-Hill Book Co., 1978.

33. G. G. Skitek, and S. G. Marshall, "Electro-magnetic Concepts and Applications" Prentice Hall, Inc., 1982.

34. M. S. Sarma, *Electric Machines: Steady-State Theory and Dynamic Performance,* Wm. C. Brown Publishers, 1985.

35. G. J. Thaler, and M. L. Wilcox, *Electric Machines: Dynamics and Steady State,* John Wiley & Sons, Inc., 1966.

36. H. H. Woodson, and J. R. Melcher, *Electro-mechanical Dynamics,* part I, "Discrete Systems," John Wiley & Sons, Inc., 1968.

37. *Technical Data on ALNICO v-7,* the Arnold Engineering Co., Bulletin PM-123, Feb. 1963.

Chapter
4

Systems of Electromagnetic Energy Capture

In this chapter, conceptual review of conventional electromagnetic systems for energy storage will be presented, including air-cored solenoids, condensers having minimum dielectric loss, and combination of coils and capacitances. The purpose of these energy storage devices or systems is the capture of energy content from high-voltage switching surges (whether intentional or accidental) and lightning surges. After that, in a planned sequence, the stored energy will be released to recharge power-producing generators such as storage batteries and/or the redox flow cells, where the stored energy could also be used to produce hydrogen and oxygen by applying it through the process of electrolysis on water (which could be either in liquid or vapor states). Upon successful separation of H_2 and O_2, they would be used as the fuel and oxidizer in the process of performance of the $H_2 - O_2$ fuel cell.

In addition, the captured electric energy could be used in another function of electrolysis and electroseparation to release hydrogen and oxygen from material of prepared refuse in the form of a complex hydrocarbon, followed subsequently by powering the generating action of the bioelectrochemical fuel cell.

4.1   Energy Storage by Inductor

In Fig. 4.1 is an air-cored solenoid having a number of turns N, through which a current is impressed.

From Faraday's law, we can write:

$$e = -L \frac{di}{dt} \text{ volts} \qquad (4.1)$$

137

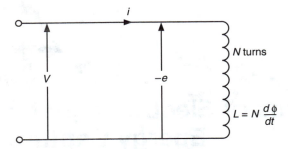

Figure 4.1 Air-cored inductor magnetic flux ϕ.

where e = the induced voltage
L = the coil self-inductance measured in henries. Then

$$p = \frac{dw}{dt} = -Li\,\frac{di}{dt}\ \text{watts}$$

Therefore,

$$W_{mag} = -L\int_0^I i\,\frac{di}{dt}\,dt$$

$$= -\frac{1}{2}\,LI^2\ \text{joules} \qquad (4.2)$$

I = the final value of current buildup in the solenoid

Usually, $i(t)$ with zero current in the solenoid at $t = 0^+$ (i.e., at the moment the current surge entry in the solenoid) is expressed by:

$$i(t) = \frac{E}{R}\,[1 - e^{-\frac{t}{\tau}}] \qquad (4.3)$$

where E = the initiating voltage surge
R = the solenoid resistance at power frequencies
τ = the time-constant of the solenoid, where

$$\tau = \frac{R}{L}\ \text{seconds}$$

Here is the appropriate location to introduce the well-known maximum-flux linkage theorem, which states that the current induced in a resistance-free coil could remain in existence indefinitely, even after the source had ceased.

This is shown following. Rewriting Eq. (4.1) with $v(t) = -e(t) = 0$, where $v(t)$ is the initiating voltage surge:

$$L\frac{di}{dt} = 0 = N\frac{d\phi}{dt} \tag{4.4}$$

$$\frac{d\phi}{dt} = 0 = \frac{di}{dt} \tag{4.5}$$

From Eqs. (4.4) and (4.5), we can write:

ϕ = a constant and also

I = a constant

This means the current I and Q, each representing the finally settled values for the electric current and the magnetic flux linkage, will remain stored in the coil core nonmagnetic medium indefinitely.

The other fact which must be demonstrated is that nonmagnetic solenoid medium will result in maximizing the energy stored as shown:

$$e = -L\frac{di}{dt}$$

$$= -N\frac{d\phi}{dt}$$

$$= -N\frac{dBA}{dt}$$

And, if A is to remain constant, therefore,

$$e = -NA\frac{dB}{dt} \tag{4.6}$$

For a solenoid so long that its ratio of axial length to its diameter is greater than 10, the magnetic field intensity vector $\bar{H}$ is expressed by:

$$H \simeq \frac{Ni}{l}$$

or

$$i = \frac{Hl}{N} \tag{4.7}$$

Therefore, from Eqs. (4.6) and (4.7), it follows that

$$\frac{dw}{dt} = NA\frac{dB}{dt}\frac{H\rho}{N}$$

Since in anisotropic medium $B = \mu H$, where μ is the scalar magnetic permeability, the time rate of change of magnetic energy W_m becomes:

$$\frac{dW_m}{dt} = lA\mu H \frac{dH}{dt}$$

and hence

$$W_{mag} = lA\mu \frac{H^2}{2} \text{ joules} \tag{4.8}$$

$$= lA \frac{B^2}{2\mu} \text{ joules} \tag{4.9}$$

where lA is the solenoid medium volume. We can write as follows the expression for the volume density of magnetic energy stored for a non-magnetic medium where the magnetic permeability is smallest and usually identified by μ_0.

Where $\mu_0 = 4\pi \times 10^{-7}$ henry/meter

$$W_{mag}/\text{volume} = \frac{B^2}{2\mu_0} \text{ joules/m}^3 \tag{4.10}$$

It is important to note the inertia property of the inductor, which states that the value of electric current in the inductor cannot change between the moments of O^+ and O^-, i.e., immediately before and after the initiation of a disturbance.

4.1.1 Overhead transmission line

Considering a point (x, y, z) located on a transmission line, this author developed a solution for the induced magnetic flux density vector $\bar{B}$ using the method of magnetic moment, and published it in his book, *High Voltage Engineering in Power Systems* in 1992 (CRC Press, Boca Raton, Florida).

That equation is given following. The $\bar{B}$ field was expressed by:

$$\bar{B} = \mu_0 \frac{j4}{c} Q \sum_{n,m=1}^{\infty} \sum_{n,m=1}^{\infty} \left(\frac{e^{-jX_{n,m}}}{X_{n,m}^2} + j\frac{e^{-jX_{n,m}}}{X_{n,m}} \right) (\hat{a}_y - \hat{a}_z)$$

$$\int \left[\frac{\delta^2(t)}{c} \cos ct\, dt + \frac{3\delta(t)}{c^2} \sin ct\, dt - \frac{2\delta(t)}{c^3} \sin ct\, dt \right.$$

$$\left. - \frac{1}{c^3} \delta''(t - \tau)\sin ct\, dt \right] \tag{4.11}$$

where $Q = f(J_c) = kJ_c$, k is a constant

$J_{c,v}$ = the incident induced current surge (conductive and convective)

c = velocity of light

τ = time-delay parameter

$\delta(t)$ = a unity impulse and $\delta''(t)$ is the triplet impulse

Therefore, magnetic energy density stored in the nonmagnetic core of the solenoid could be expressed by the square of the $|\bar{B}|$ field induced divided by $2\mu_0$.

To account for the corresponding current in the storing inductor, the current density vector J_c is expressed from Eq. (4.3), whereby

$$J_{c-coil}(t) = \frac{E}{RA}\left[1 - e^{-\frac{Rt}{L}}\right] \tag{4.12}$$

where A is the coil-line cross section.

Therefore, the parameter Q in Eq. (4.11) is:

$$Q_{c,v} = K\frac{E_{c,v}}{RA}\left[1 - e^{-\frac{Rt}{L}}\right] \tag{4.13}$$

4.1.2 Transformer

In a book published by CRC Press in 1992 entitled *High Voltage Engineering in Power Systems,* this author calculated the current induced in a transformer due to a lightning surge. They are reproduced here:

$$i_{transformer}(t) = \frac{1}{L}\left\{A_2 U_{-1}(t) + A_3(A_c - A_v)\left[\frac{t^2}{2} - tU_{-1}(t - t_1)\right]\right\}$$

$$[J_{cL}(t) - J_{vL}(t)]\sqrt{\frac{j}{\pi}}\frac{1}{R\hat{y}}\sum_{n=0}^{\infty}\frac{j^n}{2^n n!}$$

$$\left[\frac{X^{n+\frac{1}{2}}e^{-jnx}}{jn^2 R} - \frac{n+\frac{1}{2}}{jn^2 R}\int x^{n-\frac{1}{2}}e^{-jnx}\,dx\right.$$

$$+ \frac{x^{n+\frac{1}{2}}e^{-jnx}}{j^n}\frac{n-1}{j^n}\int x^{n-2}\,d^{-jnx}\,dx$$

$$\left. + \frac{j^{2n-1}(n-1)x^{\frac{3}{2}}}{32}\right] \tag{4.14}$$

where $R = R_c$ for conductive stroke
$\quad\quad\quad = R_v$ for conductive stroke

$$J_{cL}(t) = A_c t, \ 0 < t < t_1$$
$$= A_c(t - t_1)U_{-1}(t - t_1), \ t > t_1$$
$$J_{vL}(t) = A_v t. \ 0 < t < t_1$$
$$= A_v(t - t_1)U_{-1}(t - t_1), \ t > t_1$$

Current flow in the storing inductor placed at the entry of a transformer is, according to Eq. (4.3):

$$i_{coil}(t) = AJ_{transformer}(t)\left[1 - e^{-\frac{Rt}{L}}\right] \tag{4.15}$$

Again, A is the cross section of the inductor line.

The energy stored in the inductor produced by i_{coil} is given by:

$$W_{mag}(t) = \frac{1}{2}\, Li_{coil}^2 \text{ joules} \tag{4.16}$$

For an inductor with very small resistance, the magnetic energy stored, expressed by Eq. (4.16), is the maximum which will remain stored in the inductor nonmagnetic medium for a very long time.

4.2   Energy Stored in Capacitor

A voltage surge V_s applied across a capacitor will charge it to $q(t)$ coulombs. Resistance (r) represents the dielectric loss for a leaky capacitor. For an ideal capacitor, $r \to \infty$. This is shown in Fig. 4.2.

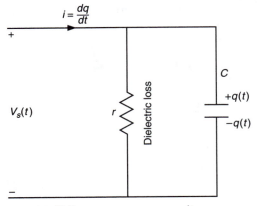

Figure 4.2 Voltage surge across capacitor.

If, at $t = 0^+$, which is the moment immediately after closing the switch (s), the voltage across the capacitor is expressed by:

$$V_c(t) = [1 - e^{-\frac{t}{\tau}}]V_s \qquad (4.17)$$

where $\tau = rc$ seconds.

We have to remember the inertia property for the capacitance; namely, the voltage across a capacitor remains constant before and after a sudden change, such as at the moment of 0^- before closing a switch and at a moment of 0^+ after closing a switch.

For an ideal capacitance $r \to \infty$ (negligible dielectric loss), and hence the voltage V_s, will set at the two terminals of the two plates immediately at $t = 0^+$.

To find the energy stored in a red capacitance, we can write the following:

$$i = c \, \frac{dV_c(t)}{dt} \qquad (4.18)$$

Electric energy W_e is therefore given by:

$$W_e(t) = \int c \, \frac{dV_c}{dt} \, V(t)dt$$

$$= \frac{1}{2} \, CV_c^2(t) \text{ joules} \qquad (4.19)$$

Since the capacitance c is expressed by:

$$c = e \, \frac{A}{d} \qquad (4.20)$$

where e is the capacitance permitivity constant, A, d is the area of each capacitance plate and the separation between the plates, respectively.

For free space, $e = e_0 = 8.85 \times 10^{-12}$ farad/meter. Therefore,

$$W_e(t) = \frac{1}{2} \, \frac{eA}{d} \, V_c^2(t) \text{ joules} \qquad (4.21)$$

A large value for e will enhance tremendously energy storage by the capacitance.

Q is the electric charge on each conducting plate, and since

$$Q(t) = C \, V(t) \qquad \text{coulombs}$$

therefore,

$$W_e(t) = \frac{1}{2} \, \frac{eA}{d} \times \frac{e^2A^2}{d^2} \, Q^2(t) \text{ joules}$$

The electric energy density W_{ed} is

$$W_{ed}(t) = \frac{W_{ed}}{Ad}$$

$$= \frac{1}{2} \frac{eA}{d} \frac{1}{Ad} V_c^2(t)$$

$$= \frac{1}{2} \frac{e}{d^2} V_c^2(t) \text{ joules/m}^3 \tag{4.22}$$

If the electric field is uniformly distributed across the two parallel plates, i.e.,

$$E(t) = \frac{V(t)}{d}$$

therefore,

$$W_{ed} = \frac{1}{2} \frac{e}{d^2} E_c^2(t)d^2$$

$$= \frac{1}{2} eE_c^2(t)$$

$$= \frac{1}{2} eD^2(t) \text{ joules/m}^3 \tag{4.23}$$

For a capacitor with initial voltage $V_0(t = 0^-)$, $V_c(t)$ is given by:

$$V_c(t) = V_s + [V_0 - V_s] e^{-\frac{t}{RC}} \tag{4.24}$$

4.2.1 Lightning surge on a transmission line

In *High Voltage Engineering in Power Systems* (CRC Press, 1992) this author developed a solution for the induced voltage on an overhead transmission line due to a sharp and sustained lightning surge, covering the conductive as well as the convective strokes. The solution of $V_{induced}(x,t)$ is reproduced here:

$$V_{induced}(x,t) = \frac{2C^2}{V_i(x = t)U_{-1}(t - t_0)} \sqrt{\frac{j}{\pi}} \frac{1}{\hat{y}}$$

$$\left\{ \frac{A_c}{R_c} \sum_{n=0}^{\infty} \frac{j^n}{2^n n!} \left[\frac{\rho^{n-\frac{1}{2}} e^{-jn\rho}}{jn^2 R_c} - \frac{n + 1/2}{jn^2 R_c} \int \rho^{n-\frac{1}{2}} e^{-jn\rho} d\rho \right. \right.$$

$$\left. \left. + \frac{\rho^{n-1} e^{-jn\rho}}{jn} - \frac{n - 1}{jn} \int \rho^{n-2} e^{-jn\rho} d\rho + \frac{j2^{n+1}(n - 1)!}{3\sqrt{2}} \rho^{\frac{3}{2}} \right] \right.$$

$$+ \frac{A_c}{R_v} \sum_{n=0}^{\infty} \frac{j^n}{2^n n!} \left[\frac{\rho^{n+\frac{1}{2}} e^{-jn\rho}}{jn^2 R_v} - \frac{n+1/2}{jn^2 R_v} \int \rho^{n-\frac{1}{2}} e^{-jn\rho} d\rho \right.$$

$$\left. + \frac{\rho^{n-1} e^{-jn\rho}}{jn} \int \rho^{n-2} e^{-jn\rho} dp + \frac{j2^{n+1}(n-1)! \rho^{\frac{3}{2}}}{3\sqrt{2}} \right] \Big\} \qquad (4.25)$$

where V_i, the inducing voltage on a transmission line due to a combination of conductive and convective strokes of a sharp, sustained lightning surge, is given in my book, *High Voltage Engineering in Power Systems,* and reproduced here:

$$V_{ic,v} = A \sqrt{\frac{j}{pi} \frac{1}{R_{c,v} \hat{y}}} \sum_{n=0}^{\infty} \frac{j^n}{2^n n!}$$

$$\times \left[\frac{x^{n+\frac{1}{2}} e^{-jnx}}{jn} - \frac{n+\frac{1}{2}}{jR_{c,v} n^2} \int x^{n-\frac{1}{2}} e^{-jnx} dx \right.$$

$$\left. + \frac{x^{n-1} e^{-jnx}}{jn} \frac{n-1}{jn} \int x^{n-2} e^{-jnx} dx + \frac{j^{n+1}(n-1)!}{3\sqrt{2}} \times \frac{3}{2} \right] \qquad (4.26)$$

where x is the horizontal projection of any point in space along the transmission line.

$$J_{c,v} = A_{c,v} U_{-1}(t) \qquad \text{for } 0 < t < t_1$$

$$= A_{c,v}(t - t_1) \qquad \text{for } t > t_1 \qquad (4.27)$$

where

$J_{c,v}$ is the conductive or convective current density

ρ is the radial distance of any point in space

c is the velocity of light

V_i is the inducing voltage at $x = l$

$\hat{y}$ is the admitivity of space $= \hat{\sigma} + jw\hat{e}$

$J_{c,v}$ is the J_c for conductive current density produced by the conductive lightning stroke and J_v for the convective current density produced by the convective lightning stroke

According to Eq. (4.23), the electric energy density that could be stored in a condenser system is given by:

$$\frac{W_e}{Ad} = \frac{1}{2} \frac{e}{d^2} V_c^2 (x, t) \text{ joules/m}^3 \qquad (4.28)$$

where $V_c(x,t)$ is the $V_c(t)$ indicated by Eq. (4.24), or:

$$V_c(x,t) = V(x,t) + [V_0 - V(x,t)]e^{-\frac{t}{RC}} \qquad (4.29)$$

Also, the V_s in Eq. (4.24) is the $V(x,t)$ and, of course, V_0 is an already existing voltage stored in the condenser system and available at $t = 0^-$, the moment preceding the incident-induced voltage due to a lightning surge induced at a point x of a transmission line.

4.2.2 Lightning voltage surge on transformer

In his book *High Voltage Engineering in Power Systems*, this author developed a solution for the voltage induced at a transformer due to a lightning surge. That solution is reproduced here:

$$V_{trans\text{-}induced}(x,\ t) = [A_2\delta(t) + A_3(A_c - A_v)t - U_{-1}(t - t_1)]\,f(x) \qquad (4.30)$$

where

$$A_2 = E_0\,\frac{\delta(t) - U_{-1}(t - t_1) - U_{-1}(t_1)/c + t}{\delta(t) - U_{-1}(t - t_1) + t} \qquad (4.31)$$

$$A_3 = \frac{E_0}{A_c - A_v} \qquad (4.32)$$

$$f(x) = \sqrt{\frac{j}{\pi}}\,\frac{1}{R_{c,v}} \sum_{n=0}^{\infty} \frac{j^n}{2^n n!}$$

$$\left[\frac{x^{n+\frac{1}{2}}\,e^{-jnx}}{jn} - \frac{n-1}{jn}\int x^{n-2}e^{-jnx}dx + \frac{j2x^{n-1}(n-1)!}{32}\right] \qquad (4.33)$$

A condenser system installed just at the entry of a transformer could act to capture the $V_{trans}(x,\ t)$ produced by conductive and convective lightning strokes. Energy density captured and stored in the condenser system is expressed as follows, using Eq. (4.22):

$$\frac{W_e(t)}{Ad} = \frac{1}{2}\,\frac{e}{d^2}\,V_{c-trans}^2\,(x,\ t) \qquad (4.34)$$

where

$$V_{c-trans}\,(x,\ t) = V_{trans\text{-}induced}\,[1 - e^{-\frac{t}{rc}}] \qquad (4.35)$$

Equation (4.34) is valid if the condenser is uncharged.

However, if there is an initial stored voltage at the condenser, then according to Eq. (4.24):

$$V_{c-trans}(x, t) = V_{trans}(x, t) + [V_0 - V_{trans}(x, t)]e^{-\frac{t}{rc}} \qquad (4.36)$$

where $V_{trans}(x, t)$ is given by Eq. (4.29).

4.3   Storage system of L and C in parallel

In reference to Fig. 4.3:

$$e_1 = e_2 \qquad (4.37)$$

or

$$e_{t_1} + e_{r_1} = e_{t_2} \qquad (4.38)$$

$$i_1 = i_2 + C \frac{de_{t_2}}{dt} + \frac{1}{L} \int e_{t_2} dt \qquad (4.39)$$

Using Equation (4.39):

$$\frac{e_{t_1}}{z_1} - \frac{e_{r_1}}{z_2} = \frac{e_{t_2}}{z_2} + C \frac{de_{t_2}}{dt} + \frac{1}{2} \int e_{t_2} dt \qquad (4.40)$$

Adding Eqs. (4.38) and (4.40) multiplied by z_1:

$$\frac{1}{L} \int e_{t_2} dt + e_{t_2} \frac{z_1 + z_2}{z_1 z_2} + C \frac{de_{t_2}}{dt} = 2 \frac{e_{t_1}}{z_1} \qquad (4.41)$$

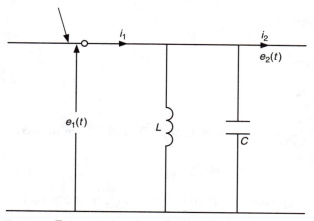

Figure 4.3   Energy storage through parallel L and C.

Examination of e_{t_1}, represented by Eqs. (4.25) and (4.26), reveals the following. Inspecting Eqs. (4.25) and (4.26), we can see that e_{t_1} for a transmission line is a step function in terms of time, given by:

$$A'U_{-1}(t)G(x)$$

To obtain a solution for e_{t_2} from Eq. (4.41), we proceed as follows:

$$\frac{e_{t_2}}{L} + \left(\frac{z_1 + z_2}{z_1 z_2}\right)\frac{de_{t_2}}{dt} + C\frac{d^2 e_{t_2}}{dt^2} = 0$$

or

$$e_{t_2} + \left(L\,\frac{z_1 + z_2}{z_1 z_2}\right)\frac{de_{t_2}}{dt} + LC\frac{d^2 e_{t_2}}{dt^2} = 0 \tag{4.42}$$

The characteristic definition of Eq. (4.42):

$$LCD^2 + L\,\frac{z_1 + z_2}{z_1 z_2}\,D + 1 = 0$$

Let $LC = a_1$ and $L\,\dfrac{z_1 + z_2}{z_1 z_2} = a_2$

Therefore,

$$D_{1,2} = \frac{-a_2 \pm \sqrt{a_2^2 - 4a_1}}{2a_1}$$

Hence,

$$e_{t_2} = k_1 e^{D_1 t} + k_2 e^{D_2 t} \tag{4.43}$$

The initial conditions on e_{t_2} are:

$t = 0^+$

$e_{t_2} = 0$

$t \to \infty$

$e_{t_2} \to 0$

which implies that $k_1 = -k_2 = K$ and that both D_1 and D_2 are negative. Therefore,

$$e_{t_2} = k\,[e^{-D_1 t} - e^{-D_2 t}] \tag{4.44}$$

Now, turning to energy stored in L and C, W_{mag} in L from Eq. (4.16):

$$= \frac{1}{2}\,LI^2_{coil}$$

where
$$I_{coil} = \frac{1}{L} \int e_{t_2} dt$$

$$= \frac{K}{L} \left[-\frac{1}{a_1} e^{-a_1 t} + \frac{1}{a_2} e^{-a_2 t} \right] + I_0 \tag{4.45}$$

Therefore,

$$W_{mag} = \frac{L}{2} \left\{ \frac{K}{L} \left[\frac{e^{-a_1 t}}{a_2} - e^{-a_1 t} a_1 \right]^2 + I_0 \right\}^2 \tag{4.46}$$

Then the energy stored in the capacitance C from Eq. (4.21):

$$W_e = \frac{1}{2} \frac{eA}{d} e_{t_2}^2$$

$$= \frac{1}{2} \frac{eA}{d} k^2 [d^{-a_1 t} - d^{-a_2 t}]^2 \text{ joules} \tag{4.47}$$

$$W_{mag - avc} = \frac{1}{2} LI_0^2 \text{ joules}$$

Design conditions on network elements of Fig. 4.3:

$$D_1 = -\frac{a_2}{2a_1} + \frac{1}{2a_1} \sqrt{a_2^2 - 4a_1}$$

$$= -\frac{Z_1 + Z_2}{2LCZ_1Z_2} + \frac{1}{aLC} \sqrt{\frac{(Z_1 + Z_2)^2}{Z_1^2 Z_2^2} - 4LC} \tag{4.48}$$

For D_1 to be negative, we observe the following:

$$\frac{Z_1 + Z_2}{2LCZ_1Z_2} > \frac{1}{2LC} \sqrt{\frac{(Z_1 + Z_2)^2}{Z_1^2 Z_2^2} - 4LC} \tag{4.49}$$

Then

$$\frac{(Z_1 + Z_2)^2}{Z_1^2 Z_2^2} > \left[\frac{(Z_1 + Z_2)^2}{Z_1^2 Z_2^2} - 4LC \right] \tag{4.50}$$

Of course for D_2 to be negative, no restriction can be imposed on L, C, Z_1, and Z_2.

4.4 Storage System Involving Series *L* and Parallel *C*

As shown in Fig. 4.4, continuity equation pertaining to currents and voltages are expressed as follows:

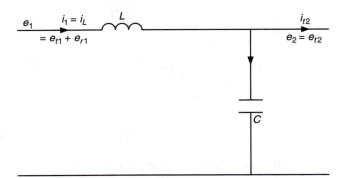

Figure 4.4 Energy storage of series L and shunt C.

$$e_{t_1} + e_{r_1} = \frac{L}{Z_1} \frac{de_{t_1}}{dt} + e_{t_2} \tag{4.51}$$

$$i_1 = i_2 + C \frac{de_{t_2}}{dt} \tag{4.52}$$

Then

$$\frac{e_{t_1}}{Z_1} - \frac{e_{r_1}}{Z_1} = \frac{e_{t_2}}{Z_2} + C \frac{de_{t_2}}{dt} \tag{4.53}$$

or

$$e_{t_1} - e_{r_1} = \frac{Z_1}{Z_2} e_{t_2} + cZ_1 \frac{de_{t_2}}{dt} \tag{4.54}$$

Adding Eqs. (4.51) and (4.54) results in:

$$2e_{t_1} = e_{t_2} \left(1 + \frac{Z_1}{Z_2} \right) + CZ_1 \frac{de_{t_2}}{dt} + \frac{L}{Z_1} \frac{de_{t_1}}{dt} \tag{4.55}$$

From Sec. 4.3, we indicated that e_{t_1}, which is the voltage induced on a transmission line due to an ideal lightning current surge (as a step function), is also a step multiplied by a function of space.

e_{t_1} for a transmission line is a step function $= A'U_{-1}(t)$.

Therefore, from Eq. (4.55), we can write:

$$2A'U_{-1}(t) - \frac{L}{Z_1} U_0(t) = CZ_1 \frac{de_{t_2}}{dt} + \left(1 + \frac{Z_1}{Z_2} \right) e_{t_2} \tag{4.56}$$

or taking the Laplace transform of the preceding equation:

$$\frac{2A'}{S} - \frac{L}{Z} = CZ_1SE_{t_2}(s) + \left(1 + \frac{Z_1}{Z_2}\right)E_{t_2}(s)$$

$$\frac{2AZ_1 - SL}{SZ_1} = E_{t_2}(s)\left[SCZ_1 + 1 + \frac{Z_1}{Z_2}\right] \tag{4.57}$$

Therefore,

$$E_{t_2}(s) = \frac{2AZ_1 - SL}{SZ_1} \times \frac{Z_2}{Z_1 + Z_2 + SCZ_1Z_2}$$

$$= \frac{-Z_2L\left(S - \dfrac{2A'Z_1}{L}\right)}{SZ_1^2Z_2C\left(S + \dfrac{Z_1 + Z_2}{CZ_1Z_2}\right)} \tag{4.58}$$

Taking the inverse transform of Eq. (4.58) gives:

$$e_{t_2}(t) = \frac{2A'}{Z_1 + Z_2} - \left(\frac{L}{CZ_1^2} + \frac{2A'}{Z_1 + Z_2}\right)e^{-\frac{z_1 + z_2}{cz_1z_2}t} \tag{4.59}$$

Energy stored in the condenser (initially assumed to be uncharged) is, using Eq. (4.21), as follows:

$$W_e(t) = \frac{1}{2}\frac{eA}{d}\left[\frac{2A'}{Z_1 + Z_2} - \left(\frac{L}{CZ_1^2} + \frac{2A'}{Z_1 + Z_2}\right)e^{-\frac{z_1 + z_2}{cz_1z_2}t}\right]^2 \text{ joules} \quad (4.60)$$

To find W_{mag} stored in the inductor, first we find the current i_1 from Eq. (4.52):

$$i_L = \frac{e_{t_2}}{Z_2} + C\frac{de_{t_2}}{dt}$$

$$= \frac{2A'}{Z_2(Z_1 + Z_2)} + \left[\frac{L}{CZ_1^2} + \frac{2A'}{Z_1 + Z_2}\right]\left(\frac{1}{Z_1}\right)e^{-\frac{z_1 + z_2}{cz_1z_2}} \tag{4.61}$$

W_{mag} stored in the inductor $= \tfrac{1}{2}Li_1^2$ joules. Therefore,

$$W_{mag} = \frac{L}{2}\left[\frac{2A'}{Z_2(Z_1 + Z_2)} + \left(\frac{L}{CZ_1^2} + \frac{2A'}{Z_1 + Z_2}\right)\left(\frac{1}{Z_1}\right)e^{-\frac{z_1 + z_2}{cz_1z_2}t}\right]^2 \text{ joules}$$

$$\tag{4.62}$$

W_{e-ave} is expressed as follows by inspection from Eq. (4.62):

$$W_{e-ave} = \frac{eAA'}{d(Z_1 + Z_2)} \text{ joules} \tag{4.63}$$

from which $W_{mag-ave}$ is similarly written as:

$$W_{mag-ave} = \frac{LA'}{Z_2(Z_1 + Z_2)} \text{ joules} \tag{4.64}$$

Equations (4.63) and (4.64) do not impose restrictions on L, C, Z_1, and Z_2.

4.5   Storage System Involving Parallel L and C between Z_1 and Z_2

Equations of continuity for the system shown in Fig. 4.5 are:

$$e_1 = e_2 + L\frac{di_L}{dt} = e_2 + \frac{1}{c}\int i_c dt \tag{4.65}$$

$$e_{t_1} + e_{r_1} = e_{t_2} + L\frac{di_L}{dt} = e_{t_2} + \frac{1}{c}\int i_c dt \tag{4.66}$$

$$i_1 = i_2 + i_c = i_2 \tag{4.67}$$

$$\frac{e_{t_1}}{Z_1} - \frac{e_{r_1}}{Z_1} = \frac{e_{t_2}}{Z_2} \tag{4.68}$$

$$e_{t_1} - e_{r_1} = \frac{Z_1}{Z_2} e_{t_2} \tag{4.69}$$

Adding Eqs. (4.66) and (4.69):

$$2e_{t_1} = e_2 + L\frac{di_L}{dt} + \frac{Z_1}{Z_2} e_{t_2} \tag{4.70}$$

$$= e_{t_2} + \frac{1}{c}\int i_c dt + \frac{Z_1}{Z_2} e_{t_2} \tag{4.71}$$

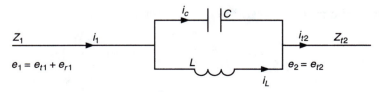

Figure 4.5   Energy storage of parallel L and C.

Also, from Eq. (4.67):

$$\frac{e_{t_2}}{Z_2} = i_2 + i_c \tag{4.72}$$

and

$$\frac{1}{c}\int i_c dt = L\frac{di_2}{dt} = V_c = V_L \tag{4.73}$$

Transforming Eqs. (4.70) through (4.73) into Laplace transforms yields:

$$E_{t_2}(s) = Z_2 I_2(s) + Z_2 I_c(s)$$

$$SLI_c(s) = \frac{I_c(s)}{SC} \quad \text{and}$$

$$2E_{t_1}(s) = SLI_L(s) + E_{t_2}(s) + \frac{Z_1}{Z_2}E_{t_2}(s) \tag{4.74}$$

where $e_{t_1}(t) = A'v_{-1}(t)G(x)$ for a transmission line.

From the set of Eq. (4.74), by taking the inverse Laplace transform, $i_L(t)$ is given as:

$$i_L(t) = \frac{[(S_1 - S_2) + S_2 e^{s_1 t} - S_1 e^{s_2 t}]2A'}{S_1 S_2 (S_1 - S_2)} \tag{4.75}$$

where

$$S_{1,2} = \frac{-a_2 \pm \sqrt{a_2^2 - 4a_1 a_3}}{2a_1} \tag{4.76}$$

$$a_1 = LCZ_2(Z_1 + Z_2)$$

$$a_2 = L$$

$$a_3 = Z_2(Z_1 + Z_2) \tag{4.77}$$

Therefore,

$$W_{mag} = \frac{1}{2}Li_2^2 \text{ joules}$$

To obtain a solution for the energy stored in the condenser, we note that

$$V_c = V_L = L\frac{di_2}{dt}$$

Therefore, from Eq. (4.62):

$$V_L(t) = V_c(t)$$

$$= L \frac{d}{dt} \left[\frac{(s_1s_2) + S_2 e^{s_1 t} - S_1 e^{s_2 t}}{S_1 S_2 (S_1 - S_2)} \right] 2A'$$

$$= \frac{2LA'}{S_1 S_2 (S_1 - S_2)} [S_1 S_2 e^{s_1 t} - S_1 S_2 e^{s_2 t}] \text{ volts} \qquad (4.78)$$

$$W_e(t) = \frac{1}{2} C V_c^2(t)$$

$$= \frac{CL^2 A'^2}{(S_1 - S_2)^2} [e^{s_1 t} - e^{s_2 t}]^2 \text{ joules} \qquad (4.79)$$

However, $W_{e-ave} \to 0$, since physical realization for $i_2(t)$ requires S_1 and S_2 to be negative.

But $W_{mag-ave}$, using Eq. (4.75), is given by:

$$W_{mag-ave} = \frac{L}{2} \left(\frac{2A'}{S_1 S_2} \right)^2$$

$$= 2L \left(\frac{A'}{S_1 S_2} \right)^2 \text{ joules} \qquad (4.80)$$

Design condition from S_1 and S_2:

$$S_{1,2} = -\frac{a_2}{2a_1} + \sqrt{a^2 - 4a_1 a_3}$$

$$= -\frac{L}{2C \, LZ_2(Z_1 + Z_2)} \pm \sqrt{L^2 - 4LCZ_2^2 (Z_1 + Z_2)^2} \qquad (4.81)$$

For S_1 to be negative

$$\frac{L}{2C \, LZ_2(Z_1 + Z_2)} > \sqrt{L^2 - 4LCZ_2^2 (Z_1 + Z_2)^2} \qquad (4.82)$$

or

$$\frac{L^2}{4L^2 C^2 Z_2^2 (Z_1 + Z_2)^2} > [L^2 - 4LCZ_2^2 (Z_1 + Z_2)^2] \qquad (4.83)$$

Then

$$\frac{1}{4C^2 Z_2^2 (Z_1 + Z_2)^2} > [L^2 - 4LCZ_2^2 (Z_1 + Z_2)^2] \qquad (4.84)$$

For S₂ to be negative. We can observe from Eq. (4.70) that restriction can be put on the elements of L, C, Z_1, and Z_2.

Therefore, the inequality of Eq. (4.84) gives the only reliable design condition.

4.6 Storage System Involving *L* and *C*

As shown in Fig. 4.6, circuit theory equations of continuity renders the following relationships:

$$i_1 = i_2 + C \, \frac{de_{t_1}}{dt} \tag{4.85}$$

$$e_1 = e_2 + L \, \frac{di_2}{dt} \tag{4.86}$$

or

$$\frac{e_{t_1}}{Z_1} - \frac{e_{r_1}}{Z_1} = \frac{e_{t_2}}{Z_2} + C \, \frac{de_{t_1}}{dt} \tag{4.87}$$

$$e_{t_1} - e_{r_1} = e_{t_2} \frac{Z_1}{Z_2} + CZ_1 \, \frac{de_{t_1}}{dt} \tag{4.88}$$

$$e_{t_1} + e_{r_1} = e_{t_2} + \frac{L}{Z_2} \, \frac{de_{t_2}}{dt} \tag{4.89}$$

Adding Eqs. (4.88) and (4.89) yields:

$$2e_{t_1} = e_{t_2} + \frac{L}{Z_2} \, \frac{de_{t_2}}{dt} + e_{t_2} \frac{Z_1}{Z_2} + CZ_1 \, \frac{de_{t_1}}{dt} \tag{4.90}$$

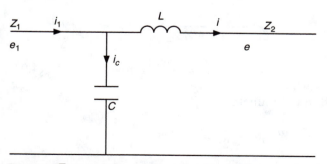

Figure 4.6 Energy storage of shunt C and series L.

Taking the Laplace transform of Eq. (4.90) with zero initial conditions gives:

$$2\,\frac{A'}{S} = E_{t_2}(s) + \frac{SL}{Z_2}\,E_{t_2}(s) + \frac{Z_1}{Z_2}\,E_{t_2}(s) + CZ_1A'$$

$$= E_{t_2}(s)\left[1 + \frac{Z_1}{Z_2} + \frac{SL}{Z_2}\right] + CZ_1A' \tag{4.91}$$

where e_{t_1} is considered a step function for a transmission line subjected to a sustained lightning surge and

$$= A'U_{-1}(t)G(x)$$

$$E_{t_2}(s) = \frac{2A'}{S\left[CZ_1A' + 1 + \dfrac{Z_1}{Z_2} + \dfrac{SL}{Z_2}\right]}$$

$$= \frac{2A'Z_2}{S\,[(Z_1 + Z_2 + CA'Z_1Z_2)/L + S]L} \tag{4.92}$$

Taking the inverse Laplace transform of Eq. (4.92) yields:

$$e_{t_2}(t) = \frac{2A'Z_2L}{Z_1 + Z_2 + CA'Z_1Z_2}$$

$$\times\left[1 - e^{-\frac{Z_1 + Z_2 + CA'Z_1Z_2}{L}t}\right]\text{ volts} \tag{4.93}$$

To find energy stored in the inductor, we need to find the voltage induced in L:

$$i_2 = \frac{e_{t_2}}{Z_2}$$

$$= \frac{2A'L}{Z_1 + Z_2 + CA'Z_1Z_2}$$

$$\times\left[1 - e^{-\frac{Z_1 + Z_2 + CA'Z_1Z_2}{L}t}\right]\text{ amperes} \tag{4.94}$$

Then energy stored in the inductor $W_{mag}(t)$ is:

$$W_{mag}(t) = \frac{1}{2}\,Li_2^2\text{ joules}$$

where $i_2(t)$ is given by Eq. (4.94).

The energy stored in the capacitor $W_e(t)$ is given by:

$$W_e(t) = \frac{1}{2} Ce_1^2(t) \text{ joules} \qquad (4.95)$$

where $e_1(t)$ is given by the expression $e_{t_1} + e_{r_1}$.

Therefore, average energy stored in the capacitor is given by:

$$W_{e-ave} = \frac{1}{2} C[e_1]^2 = \frac{1}{2} C[2e_{t_1}]^2, \text{ since } e_1 = 2e_{t_1}$$

$$= 2CA'v_{-1}(t)G^2(x) \text{ joules}$$

$$\text{(for a transmission line)} \qquad (4.96)$$

The average W_{mag} stored in the inductor is given by:

$$W_{mag-ave} = \frac{1}{2} Li_{2-ave}^2$$

$$= \frac{A'^2L^3G^2(x)}{(Z_1 + Z_2 + CA'Z_1Z_2)^2} \text{ joules} \qquad (4.97)$$

Equations (4.86) and (4.97) apparently do not impose any physical design condition on the elements L, C, Z_1, and Z_2.

In Secs. 4.3 through 4.6, computations for expressions regarding energy storage have been confined to voltage induced on transmission lines, while Secs. 3.1 and 3.2 dealt with voltage induced on transformers and transmission lines.

Calculations for energy stored in the L-C networks indicated in Secs. 4.3 through 4.6 are left as problems.

Also, the author has to accertain that in all network configurations presented in Secs. 4.1 through 4.6 the voltage arrester could be connected in series or parallel to limit the voltage and/or current surge allowed to penetrate the L or C element. The arrestor in question may have a limiting voltage e_0 or limiting current I_0.

4.7 The Induction Generator

If the slip of the polyphase induction motor becomes negative, where the rotor speed is larger than the synchronous speed, the rotor total resistance becomes negative $(R_{r-total} = R_r/-S)$. This implies that the stator absorbs negative power from the stator, or, in effect, supplies real power to the stator.

The induction generator as a power source (if it can be accomplished) has an advantage over the synchronous generator in that no separate DC source is needed for armature excitation, and no brushes are necessary to carry the current to the rotor or armature (the usual name for synchronous generator). Theoretically speaking, the induction generator can operate over a wide range of speed while connected to a bus-bar in an existing power system, thus eliminating the need for synchronous gear. The induction generator poses no stability problem since it cannot fall out of step as the synchronous generator.

All the foregoing remarks about the induction generator seem on the positive side; the main drawback is the need to provide prerequisite reactive power to establish the rotating mmf in the air gap. An open-circuited stator cannot supply that.

Therefore, the induction generator can operate in conjunction with synchronous motors or generators to receive from the three-phase line the necessary reactive power required to establish the exciting mmf. Also, we have to emphasize the fact that the induction generator supplies real power at a leading power factor close to unity.

The solution for the excitation problem of the induction generator to operate independently is to connect a capacitor of high-quality factor in a three-phase system across the stator, where the charged condenser will supply the initial need of the reactive power to provide the magnetizing current i_m.

This topic is being mentioned in this chapter, where several systems have been presented in connection with the modes of electromagnetic energy storage. The stored energy is basically induced by switching surges and, in particular, lightning surges.

A pinpointing application of the stored electromagnetic energy is its utilization in directly supplying the initial reactive power to operate the induction generator independently, without resort to linkage to a network containing synchronous motors or generators.

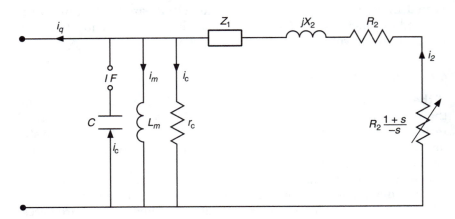

Figure 4.7 Energy stored in capacitor feeds i_m in the induction generator

One possible system for the operation of an induction generator is the one shown in Fig. 4.7, a simple storage system involving a capacitor having a stored energy W_e and connected across the stator through a centrifugal switch. At the moment of starting the induction generator, the switch (F) will release a reactive current in the three phases of the generator stator, thereby creating the rotating mmf in the air gap. Then, when the generator attains the steady state of producing the almost real power, the switch (F) will open, disconnecting the charged condenser from the stator.

$V_c(x, t)$ induced at the transformer is given by Eq. (4.33)

$V_c(x, t)$ could come from the voltage induced at a transmission line which is given by Eq. (4.25).

The system shown in Fig. 4.8 is given subsequently in the form of a single-phase equivalent circuit as shown in Fig. 4.9.

4.8 Problems

4.1 In reference to Eq. (4.11) for induced $B(x, t)$ at a point of transmission line, using the differential form of Maxwell's field equations, obtain the expression for the rotation of the induced electric field intensity vector.

4.2 Also using Eq. (4.11) for the induced $B(x, t)$ at a point of transmission line, obtain the expression for the total induced current density vector by using the differential form of Maxwell's field equations.

4.3 Using the solution of the total induced current density vector obtained in Prob. 4.2, establish the expression for the net charge density.

4.4 In reference to Eq. (4.25) for the induced voltage at a point of transmission line produced by a sustained lightning surge of step function, find the expression for the induced electric field intensity vector, and then obtain the solution for the average energy density in joules/m³.

4.5 In reference to Eq. (4.14) for the current induced at transformer structure, obtain the solution for the induced net charge density using the equation of continuity of Maxwell's field equations.

4.6 Show that the ratio of stored energy density in joules/m³ for a nonmagnetic cored coil to that of a nonelectric medium condenser is of the order of 10^4.

4.7 Equation (4.14) gives the solution for the induced current at transformer due to lightning surge. Using this equation, obtain the expression for the energy stored in an air-cored capacitor installed for surge energy storage.

4.8 In reference to the system of energy storage shown in Fig. 4.3, solve for the transmitted voltage surge if the incident surge of voltage on the transformer is that given by Eq. (4.30).

4.9 From the result obtained in Prob. 4.8, establish the expression for the energy that can be stored in a free dielectric loss capacitor.

4.10 Also from the results obtained in Prob. 4.8, establish design conditions on L and C with respect to Z_1 and Z_2.

4.11 In reference to the electromagnetic energy storage shown in Fig. 4.4, solve for the transmitted voltage surge if the incident voltage on a transformer is that given by Eq. (4.30).

4.12 From the result obtained in Prob. 4.11, establish design guides for L, C, Z_1, and Z_2 that are physically realizable.

4.13 Also using the results obtained in Prob. 4.11, obtain the expression for the energy stored in a free dielectric loss capacitor. Indicate conditions for maximizing the stored energy.

4.14 From solution of the transmitted voltage surge obtained in Prob. 4.11, obtain the solution for the corresponding electric field intensity vector, and then establish the solution for the induced magnetic induction vector $\bar{B}$. Finally, write the expression for the stored magnetic energy density. You can use Maxwell's field equations.

4.15 In reference to the electromagnetic energy storage system shown in Fig. 4.5, obtain the solution for the transmitted voltage through a transformer, using the incident voltage surge given by Eq. (4.30).

4.16 From the solution of the transmitted voltage surge obtained in Prob. 4.15, establish design conditions with respect to L, C, Z_1, and Z_2 that are physically realizable.

4.17 Using the result obtained in Prob. 4.15, establish the expression for the electrostatic energy to be stored in a free dielectric condenser.

4.18 From solution of the transmitted voltage surge in the transformer secured in Prob. 4.15, obtain the solution for the induced electric field intensity vector $\bar{E}$, and then establish the solution for the magnetic induction vector $\bar{B}$. You can use Maxwell's field equations.

4.19 In reference to the energy storage system shown in Fig. 4.6, obtain the solution for the transmitted voltage surge due an incident voltage surge expressed by Eq. (4.30).

4.20 Using the result obtained in Problem 4.19, establish the expression for the energy stored in a free dielectric loss capacitor; then show conditions for maximizing it.

4.21 Using the result obtained in Prob. 4.19, establish design guide conditions on L, C, Z_1, and Z_2 that are physically realizable.

4.22 From the result obtained in Prob. 4.19, obtain the solution for the magnetically induced electric field intensity vector, and then the magnetic induction vector $\bar{B}$, followed by the magnetic energy density.

4.23 From the solution of the transmitted voltage at a transformer obtained in Prob. 4.19, establish the solution for the magnetizing current i_m that can be used to excite a three-phase induction generator. Assume a storage capacitor is connected as shown in Fig. 4.9.

4.24 From the solution of the transmitted voltage at a transformer obtained in Prob. 4.15, establish the solution for the magnetizing current i_m that can be used to excite a three-phase induction generator. Assume a storage capacitor is connected as shown in Fig. 4.9.

4.9 References

1. Denno, K., *High Voltage Engineering in Power Systems,* CRC Press, Boca Raton, Florida, 1992.
2. Guillemin, E. A., *The Mathematics of Circuit Analysis,* John Wiley & Sons, Inc., 1958.
3. Harrington, R. F., *Time-Harmonic Electromagnetic Fields,* McGraw-Hill Book Co., 1961.
4. Matsch, L. W., *Electromagnetic & Electromechanical Machines,* 2d edition, Harper & Row Publishers, Inc., 1977.
5. Rudenberg, Reinhold, *Electrical Shock Waves in Power Systems,* Harvard University Press, 1968.

Utilization of Captured Electromagnetic Energies

In this chapter, we will discuss the process of protection from voltage and current surges, commenced by the stage of capturing their electromagnetic energies by systems comprising inductor having a small resistance and high-quality factor, condenser with a medium of high dielectric permitivity, combination of series inductor with shunt capacitance, combination of series condenser and shunt inductor, series combination of shunting inductor with a condenser, and a shunt system of parallel inductor and condenser. The primary function of these systems is the immediate capture of electromagnetic energy from the induced voltage and current surges on transmission lines, cables, transformers, and towers, as well as on cascaded connection of power system components.

The second stage in the process of protection is the meaningful economically feasible system for the storage and transform of the captured surge energies. This will involve the utilization of storage batteries, the rechargeable redox flow cells, and the process of electrolysis and electroseparation. Storage batteries will transform the applied electric energy at its terminals into chemical form at the two electrodes, after which the application of captured surge energies has to follow a process of controlled storing to chemical energy. Eventually, the stored chemical energy in the accumulators (storage batteries) will be released in the operation of discharge for supplying electric DC power. The DC power, after a process of inversion to time varying output, could be used to aid turbine-driven AC generators at periods of peak power demand.

The redox flow cell is a rechargeable electrochemical generator having the redox couple, mostly of Fe and either Ti or Cr water-based solutions. Application of electric potential at the cell terminals will oxidize Fe^2 solution (ferrous ionic solution) to the Fe^3 solution (ferric ionic solution). Upon the need of receiving power from the redox flow cell, the process will be reversed, where the ferric ionic solution will be reduced to ferrous ionic solution. The other electrode chamber is usually the titanium or cromium water-based solution which undergoes a process of oxidation during discharge and reduction throughout recharging.

Another operation where the captured surge energy could be used eventually in a useful and feasible way is the process of electrolysis. In this process, the captured electromagnetic energy from surge energies will be used in an electrochemical process known as *electrolysis* to convert H_2O from its initial form as a compound to another state in the form of a mixture of hydrogen and oxygen. Upon successful separation of H_2 and O_2, they can be used as the main fuel and oxidizer for the conventional $H_2 - O_2$ fuel cell or in any separate utilization for hydrogen and oxygen.

In this chapter, the author intends to present practical analysis and process design for the storage and transform of surge energies to usable chemical forms, and thereafter for the generation of electric power.

Other areas where stored captured electromagnetic energies could be utilized is for the electromechanical induction generator in supplying the needed magnetizing current, and for providing the exciting external magnetic field for the magnetohydrodynamic induction generator.

5.1 The Storage Battery

Well-known accumulators or storage batteries are the lead-acid battery, nickel-cadmium, sodium-sulphur, the lithium-batteries, etc. Most of the storage batteries are characterized by a relatively high starting current, short lead time, a long life cycle, reasonable cost, and efficiency on the order of 70 percent.

For the lead-acid battery, the process of charging is represented by the following electrochemical equations:

$$Pb^{+2} + 2e \rightarrow Pb \quad \text{at the cathode} \tag{5.1}$$

and

$$Pb^{+2} + 6H_2O \rightarrow PbO_2 + 4H_3O^+ + 2e \quad \text{at the anode} \tag{5.2}$$

In Eq. (5.1), the $2e$ refers to the flow of charge upon the application of electric potential at the battery terminals.

Two aspects could be used regarding electricity storage: one is based on supplying energy at minimum cost, which can be measured by the conversion efficiency, and the other for supplying steady output power, which is based on the measure of maximizing power per unit of the battery electrode or per unit weight or volume.

Faraday's law for electrochemical reaction stipulates that 96.5×10^6 coulombs of electric charge will release 1 kg-mole per unit valency. An electrode element of valency equal to 5 (the amount of charge needed to displace 1 kg-mole) is given as:

$$Q = \frac{96.5 \times 10^6 \times 5}{N_0 = 6.023 \times 10^{26}} \times 8.01 \times 10^{-19} \text{ coulombs/kg} \qquad (5.3)$$

where N_0 is Avogadro's number.

Or, the stored charge density could be expressed by:

$$Q_{stored} = \frac{96.5 \times 10^6 \times n}{M} \text{ coulombs/unit weight} \qquad (5.4)$$

N_0 is the kg-mole of electrons $= 6.023 \times 10^{26}$ electrons.

Moving from the concept of stored charge density to that of the stored energy density, we note that:

$$W_{stored} \text{ in joules/kg} = Q_{stored} V_e \qquad (5.5)$$

where V_e = the electrode potential during reaction. Therefore,

$$W_{stored} = \frac{96.5 \times 10^6 n}{M} V_e \text{ joules/kg}$$

$$= \frac{26.8 n V_e}{M} \text{ kWh/kg} \qquad (5.6)$$

where n = the valency of element
     M = the equivalent weight (in kg)

We can consider the battery to be represented by a storing capacitance (C) across which a dielectric resistance R is connected as shown in Fig. 5.1.

Let $E(t)$ be the voltage applied at the battery terminal at any time (t). Fig. 5.1 could be transformed to a model into the complex frequency domain, which is the s domain (s being the Laplace variable).

In the s domain, Fig. 5.1 becomes as shown in Fig. 5.2:

$$I(s) = I_D(s) + I_c(s) \qquad (5.7)$$

$$I_D(s) = \frac{E(s)}{R_D} \qquad (5.8)$$

C = charged capacitor

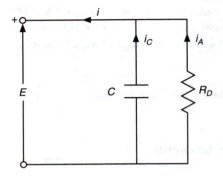

Figure 5.1 Circuit representation for the storage battery.

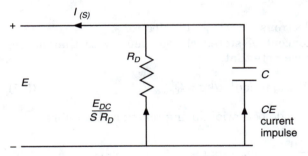

Figure 5.2 Dynamic model for the storage battery.

and
$$I_c(s) = \mathcal{L}C \, \frac{dE(t)}{dt}$$

$$= CSE(s) - CE(t = 0^+) \tag{5.9}$$

if $E(t = 0^+) = 0$. Therefore,

$$I_c(s) = CSE(s)$$

and
$$I(s) = \frac{E(s)}{R_D} + CSE(s) \tag{5.10}$$

since $E(t) = E_{DC}$ is a step function. Therefore,

$$E(s) = \frac{E_{DC}}{S}$$

and

$$I(s) = \frac{E_{DC}}{SR_D} + CS\,\frac{E_{DC}}{S}$$

$$= \frac{E_{DC}}{SR_D} + CE_{DC}$$

Hence,

$$i(t) = \frac{E_{DC}}{R_D}\,U_{-1}(t) + CE_{DC}U_0(t) \qquad (5.11)$$

where $U_{-1}(t)$ = a step function

$U_0(t)$ = an impulse function, sometimes identified by $\delta(t)$

Energy stored in the battery upon the application of $E(t)$ at its terminals could be expressed by:

$$W = \int_0^{T-\infty} E(t)i(t)dt$$

$$= \int_0^{T-\infty} E_{DC}\left[\frac{E_{DC}}{R_D}\,U_{-1}(t) + CE_{DC}U_0(t)\right]dt$$

$$= \int_0^{T-\infty} \frac{E_{DC}^2}{R_D}\,U_{-1}(t)dt + CE_{DC}^2 U_0(t)dt$$

$$= \frac{E_{DC}^2}{R_D}\,t\,\Big|_0^T + CE_{DC}^2 U_{-1}(t)\,\Big|_0^T$$

$$= \left[\frac{E_{DC}^2}{R_D}\,T + CE_{DC}^2\,U_{-1}(t)\right]\text{joules} \qquad (5.12)$$

where T is the total time of recharge in seconds.

The first term of Eq. (5.11) is the energy dissipated in the dielectric resistance R_D. If $R_D \to \infty$, the case for an ideal battery, this term becomes zero.

Therefore, the total energy stored in the ideal battery is:

$$W_e = CE_{DC}^2 U_{-1}(t),\text{ joules} \qquad (5.13)$$

From Eqs. (5.5) and (5.12), we can write

$$MW_{stored} = CE_{DC}^2 U_{-1}(t) \qquad (5.13a)$$

Therefore, total mass of a half battery participating in the ideal electrochemical reaction is given by:

$$M = [CE_{DC}^2 U_{-1}(t)/W_{stored}] \text{ kg} \qquad (5.14)$$

5.2 Controlled Chemical Storage—
Storage Batteries

In this section, controlled storage of the captured electromagnetic energy in L, C systems presented in Chap. 4 is to be introduced. Controlled process of storage involves the deposition of reactant elements with an associated electrochemical potential at either or both electrodes in the storage battery. We are going to consider selected systems design for capturing electromagnetic energy from voltage and current surges at transmission lines and transformers.

5.2.1 A series inductor
and a shunt capacitor

This system is shown in Fig. 4.4.

Steady-state voltage residing at the capturing capacitance is given by Eq. (4.59).

According to Eq. (5.13), material deposition of added electrochemical potential becomes, in the steady state:

$$M_1 = \left[\frac{C}{W_{stored}} \right] \left[\frac{2A'}{Z_1 + Z_2} \right]^2 \text{ kg} \qquad (5.15)$$

The factor $2A'/Z_1 + Z_2$ represents the steady-state final voltage buildup at the capacitance (C) in reference to Fig. 4.4.

Considering magnetic energy stored in the inductor represented by the current i_L given by Eq. (4.61), Fig. 5.3 shows the schematic for the process of controlled storing in the battery. Using Norton's theorem to transform the current source to a voltage source, we can write:

$$E(t) = i_L R_D \qquad (5.16)$$

in the steady state.

Therefore, substance deposition produced by the current i_L is

$$M_2 = \left[\frac{C}{W_{stored}} \right] \left[\frac{2A \, R_D}{Z_2(Z_1 + Z_2)} \right]^2 \text{ kg} \qquad (5.17)$$

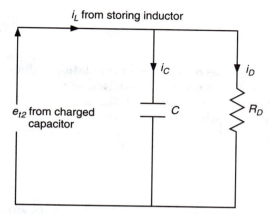

Figure 5.3 Recharging battery from either inductor or capacitor.

5.2.2 A shunt capacitance and series inductor

This system is shown in Fig. 4.6.

Subjecting the transmitted voltage $e_{t2}(t)$ given by Eq. (4.93) across a storage battery through a controlled process, the mass of deposited substance with increase in electrochemical potential is given as:

$$M_1 = \left[\frac{C_1}{W_{stored}} \right] \left[\frac{2A'Z_2L}{Z_1 + Z_2 + CA'Z_1Z_2} \right]^2 \text{kg} \qquad (5.18)$$

Equation (5.18) is valid in the steady state.

Turning to storage of captured magnetic energy stored in the inductor given by Eq. (4.8), and using the same mode discussed in Sec. 5.2.1, the mass of deposited substance with increase in electrochemical potential is given by:

$$M_2 = \left[\frac{C_1}{W_{stored}} \right] \left[\frac{2A'L}{Z_1 + Z_2 + CA'Z_1Z_2} \right]^2 \text{kg} \qquad (5.19)$$

Equation (5.19) is valid in the steady state.

Other systems involving L, C, and their combinations in a process of controlled recharging will be formulated as problems at the end of the chapter.

5.3 Controlled Chemical Storage—Redox Flow Cells

5.3.1 Discharge and recharge

The redox flow battery or cell operates on the basis of oxidation-reduction of two water-based mixtures, namely, $Fecl_3$ and $Ticl_3$ or $Fecl_3$ and $Crcl_2$. The redox couple could be a Ti or Cr solution in the anolyte, while Fe is in the catholyte. On discharge, the $Fecl_3$ will be reduced to $Fecl_2$, while the $Ticl_3$ will be oxidized to $Ticl_4$, or the $Crcl_2$ will be oxidized into $Crcl_3$.

During discharge
At the catholyte

$$Fecl_3 + \bar{e} = Fecl_2 + cl^- \qquad (5.20)$$

At the anolyte

$$Ticl_3 + cl^- = Ticl_4 + \bar{e} \qquad (5.21)$$

And the overall equation of reaction is:

$$Fecl_3 + Ticl_3 = Fecl_2 + Ticl_4 \qquad (5.22)$$

The redox flow cell is rechargeable, whereby, upon the application of electric potential at its terminals, the applied electric energy will be transformed and stored in chemical form as shown below:

During recharge
At the catholyte

$$Fecl_2 + cl^- \rightarrow Fecl_3 + \bar{e} \qquad (5.23)$$

At the anolyte

$$Ticl_4 + \bar{e} \rightarrow Ticl_3 + ci \qquad (5.24)$$

And the overall equation of reaction on recharge is:

$$Fecl_2 + Ticl_4 \rightarrow Ticl_3 + Fecl_3 \qquad (5.25)$$

Advanced designs of the redox flow cells uses $Crcl_2$ instead of $Ticl_3$ in the anolyte. Electrochemical equations in ionic states are shown following.

On discharge. Electric potential at its terminals, the applied electric energy will be transformed and stored in chemical form as shown:
At the anolyte

$$Crcl_2 - \bar{e} \rightarrow Crcl_3 + \bar{e} \qquad (5.26)$$

At the catholyte

$$Fecl_3 - \bar{e} \rightarrow Fecl_2 + cl^- \tag{5.27}$$

And the overall equation of reaction on discharge is:

$$Crcl_2 + Fecl_3 \rightarrow Crcl_3 + Fecl_2 \tag{5.28}$$

On recharge
At the anolyte

$$Crcl_3 + \bar{e} \rightarrow Crcl_2 + cl^- \tag{5.29}$$

At the catholyte

$$Fecl_2 + cl^- \rightarrow Fecl_3 + \bar{e} \tag{5.30}$$

And the overall equation of reaction on recharge is:

$$Crcl_3 + Fecl_2 \rightarrow Crcl_2 + Fecl_3 \tag{5.31}$$

The general equation of chemical reaction is written:

$$Aa + Bb = cC + dD \tag{5.32}$$

where a, b, c, and d each represent the reactant or product of reaction for the activity or relative pressure with respect to the standard, and A, B, C, and D each represent the molar concentration in a balanced cycle of reaction.

At standard pressure and temperature (SPT),

$$E_0 = -\frac{\Delta G_0}{n\Gamma} \tag{5.33}$$

where ΔG_0 = the change in Gibbs energy in joules/kg-mole at SPT
    (standard pressure and temperature)
E_0 = the voltage developed
n = the molar electrons released to an electric load
Γ = Faraday's constant (= 96.5×10^6 coulombs/kg-mole)

At any pressure and temperature P and T,

$$E = -\Delta G = -n\Gamma E \tag{5.34}$$

From Eqs. (5.21), (5.22), and (5.23),

$$E = E_0 - \frac{RT}{n\Gamma} ln - \frac{c^C d^D}{a^A b^B} \tag{5.35}$$

Focusing mainly on the process of recharge at which the captured electromagnetic energy analyzed in Chap. 4 could be transformed through a controlled process, and stored in the redox flow cell, we refer to Eqs. (5.22) and (5.25).

First, in reference to Eq. (5.22):

$$\Delta E = \frac{RT}{n\Gamma} ln \frac{P_{Ticl_3} P_{Fecl_3}}{P_{Ticl_4} P_{Fecl_2}} \tag{5.36}$$

$$= E - E_0 \tag{5.37}$$

Second, in reference to Eq. (5.22):

$$\Delta E = \frac{RT}{n\Gamma} ln \frac{P_{Crcl_2} P_{Fecl_3}}{P_{Crcl_3} P_{Fecl_2}}$$

$$= E - E_0 \tag{5.38}$$

R is the universal ideal gas constant which is equal to 8314 joules/kg - mole $- K$.

5.4   Controlled Recharge— Equivalent Capacitance

Equations (5.36) and (5.38) gave the correlation between a change in the cell internal emf developed by released electron-moles at any pressure and temperature E on one hand and that at the standard pressure and temperature E_0 on the other; n is the number of electron-moles flowing into the load in the case of discharge or an imposed or controlled flow produced by an externally applied voltage. For one mole-electron, $n = 1$, which carries a charge of 1.6×10^{-19} coulombs of electricity. However, it has been stated previously that the electric charge released or injected in any electrochemical reaction is:

$$Q_d = n\Gamma/M \text{ coulombs/kg}$$

Solving for n from Eq. (5.36), we obtain:

$$n = \frac{RT}{\Delta E \Gamma} ln \frac{P_{Ticl_3} P_{Fecl_3}}{P_{Ticl_4} P_{Fecl_2}} \tag{5.39}$$

Therefore, Q_d as the charge density becomes:

$$Q_d = \frac{RT}{\Delta E} ln \frac{P_{Ticl_3} P_{Fecl_3}}{P_{Ticl_4} P_{Fecl_2}} \text{ coulombs/kg} \tag{5.40}$$

Therefore, total charge of electricity Q to be passed in the process of recharging for conversion of $Ticl_4 - Fecl_2$ to $Ticl_3 - Fecl_3$ is accounted for by total mass M times Q_d.

A similar expression could be written for the Cr – Fe redox flow cell.

Also, we can see from Eq. (5.38) that the effective storing capacitance for the redox flow cell is given by:

$$C_{eff} = (RT)ln \ \frac{P_{Ticl_3}P_{Fecl_3}}{P_{Ticl_4}P_{Fecl_2}} \quad farads/kg \qquad (5.41)$$

For the Cr – Fe redox cell:

$$C_{eff} = (RT)ln \ \frac{P_{Crcl_2}P_{Fecl_3}}{P_{Crcl_3}P_{Fecl_2}} \quad farads/kg \qquad (5.42)$$

In Eqs. (5.41) and (5.42), T is in kelvins, M is in kg, and P_{Ticl_3}, P_{Fecl_3}, P_{Ticl_4}, P_{Fecl_2}, P_{Crcl_2}, P_{Fecl_3}, P_{Crcl_3}, and P_{Fecl_2} each represent the relative pressure with respect to one atmosphere, also known as the *activity*.

Now, since $C = e\frac{A}{d}$, we are moving to identify the effective dielectric permitivity e for the redox flow cell, which could be determined on the basis that if A were to represent the area of either the catholyte chamber or the anolyte and if d identifies the equivalent separation between the catholyte and anolyte flow surfaces, then the equivalent permitivity of the redox flow cell becomes:

$$e_{Fe - Ti} = \frac{RTd}{A} \ ln \ \frac{P_{Ticl_3}P_{Fecl_3}}{P_{Ticl_4}P_{Fecl_2}} \quad farads/meter \qquad (5.43)$$

$$e_{Fe - Cr} = \frac{RTd}{A} \ ln \ \frac{P_{Crcl_2}P_{Fecl_3}}{P_{Crcl_3}P_{Fecl_2}} \quad farads/meter \qquad (5.44)$$

5.4.1 Recharge from charged condenser

This refers to a process of controlled recharge for the redox flow cell from an already charged condenser by captured voltage or current surges located at certain preassigned stations along a transmission line close to transformer stations, line towers, and power plants. Figure 5.4a, b, c, d shows simplistic systems involving the application of the captured electromagnetic energy by a shunt capacitance or series inductor to recharge a redox flow cell. The process of controlled recharging is geared to store electromagnetic energy in the redox couple, namely, to convert chemically the $Fecl_2$ to $Fecl_3$ and $Ticl_4$ to $Ticl_3$, or the $Crcl_3$ to $Crcl_2$. Figure 5.4a represents a simple circuit to apply electric energy stored in C by the voltage induced on a transmission line due to

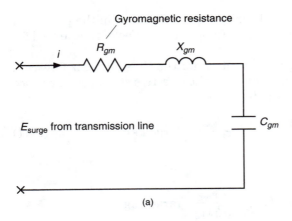

(a)

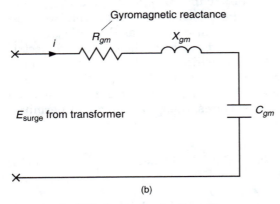

(b)

Figure 5.4 (*a, b*) Recharge of redox flow cell.

lightning or switching surges. Figure 5.4*b* represents the application of the voltage stored in *C* by the voltage induced on a transformer by lightning or switching surges.

The emf (*E*) shown in Eqs. (5.36) through (5.38) is the controlled voltage applied from the storage capacitor.

If the storage capacitor is installed at a station on a transmission line, the voltage stored in the capacitor due to an induced surge on the transmission line by a lightning surge is given by:

$$V_c(t, x) = V(x, t) + [V_0 - V(x, t)]e^{-\frac{t}{RC}} \qquad (5.45)$$

where $V(x, t)$ is given by Eq. (4.24) and V_0 is an already existing voltage across the storing capacitor. Or, if $V(x, t)$ is the voltage induced at a transformer, its solution is given by Eqs. (4.30) through (4.33).

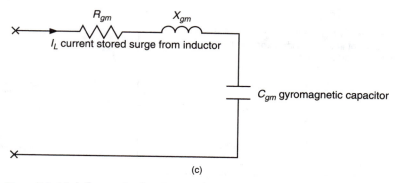

(c)

Figure 5.4 *(c)* recharge of redox flow cell at transmission line.

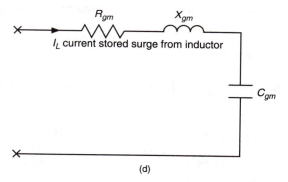

(d)

Figure 5.4 *(d)* recharge of redox flow cell at transformer.

5.4.2   Recharge from storage inductor

The emf (E) shown in Eq. (5.38) is the controlled voltage that can be derived from the magnetic energy stored in the inductor which can be installed at certain locations along a transmission line or very close to a transformer or line tower, as shown in Fig. 5.4c,d. The current induced in the inductor due to an induced magnetic induction vector expressed by Eq. (4.11) is given by Eq. (4.12). This is produced by lightning surge. The emf (E) that can be derived from the current source stored in the inductor on a transmission line is shown in Fig. 5.5, where

$$I_{inductor} = AJ \qquad (5.46)$$

$$c, \upsilon$$

$$coil$$

If the emf (E) to recharge the redox flow cell has to come from an inductor installed at a transformer, that current stored is expressed by

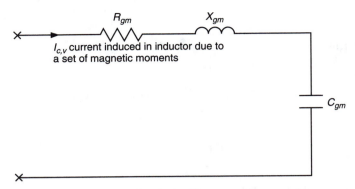

Figure 5.5 Recharge of redox flow cell by magnetic moments.

Eq. (4.12). This inductor current is produced by a lightning or switching surge.

5.4.3 A third system for controlled recharging of the redox flow cell

A third system for controlled recharging of the redox flow cell is the system shown in Fig. 5.6.

The emf (E) indicated in Eq. (5.36) will come from a process of controlled recharging of e_{t2} given by Eq. (4.93), where in the steady state:

$$E_{v_2} = \frac{2A'Z_2L}{Z_1 + Z_2 + CA'Z_1Z_2} \tag{5.47}$$

where the charge density to pass for liberating one kg of $Fecl_3$ and either $Ticl_3$ or $Crcl_2$ is:

$$Q_d = \frac{RT(Z_1 + Z_2 + CA'Z_1Z_2)}{2A'Z_2L} \, ln \, \frac{P_{Ricl_3}P_{Fecl_3}}{P_{Ticl_4}P_{Fecl_2}} \tag{5.48}$$

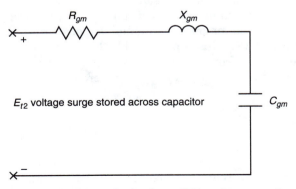

Figure 5.6 Recharge of redox flow cell from shunt capacitor-series inductor.

Recharging processes involving other L, C combinations presented in Chap. 4 are left as problems.

5.5   Production of H_2 and O_2 Mixture (Electrolysis and Electroseparation)

H_2 and O_2 are the fuel and the oxidizer for the conventional hydrogen fuel cell shown in Fig. 5.7. The ideal equation of reaction is:
   At the cathode

$$2H_2 \rightarrow 4H^+ + 4\bar{e} \tag{5.49}$$

At the anode

$$O_2 + 4H^+ + 4\bar{e} \rightarrow H_2O \tag{5.50}$$

The overall equation of reaction is:

$$2H_2 + O_2 \rightarrow 2H_2O \tag{5.51}$$

The actual equation of reaction involves heat release, electrical energy release, and mechanical work absorbed as shown:

$$2H_2 + O_2 = 2H_2O + 4eE + W_m - 3RT \tag{5.52}$$

where $3RT$ = mechanical work absorbed for pumping out any accumulated liquid H_2O
   $4eE$ = electric energy released in joules/kg-mole
   W_m = heat energy released in joules/kg-mole
   E = internally developed emf by the fuel cell that can be calculated from the Nernst equation:

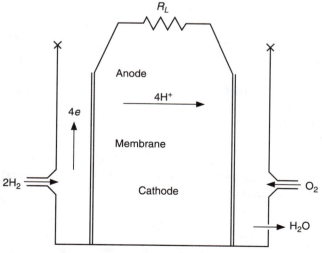

Figure 5.7 Hydrogen-oxygen fuel cell.

$$E = E_0 - \frac{RT}{n\Gamma} \ln \frac{P_C^c P_D^d}{P_A^a P_B^b} \qquad (5.53)$$

where n = the number of electron-moles released to the external circuit, which equals 4 in the present case

Γ = Faraday's constant (= 96.5×10^6 coulombs/kg-mole)

P_C, P_D = the activities or relative pressures of the products, which is H_2O in this case

P_A, P_A = the activities or relative pressures of the reactants, which are H_2 and O_2

a, b = the molar concentration for the reactants; in this case, $a = 2$ and $b = 1$

c, d = the molar concentrations for the products; in this case, $c = 2$ and $d = 0$

The process of electrolysis is the reverse of the process of electrochemical reaction as indicated by Eq. (5.52). The passage of $(4e)(E)$ joules/kg-mole will lead to the release of two moles of hydrogen and one mole of oxygen.

However, the process also involves the supply of molecular heat to account for W_m and a pumping process of H_2O if it is in liquid state. The release of H_2 and O_2 throughout the process of electrolysis is in the form of mixture, and therefore another cycle of electroseparation is needed to secure H_2 and O_2 for reuse in performance of conventional fuel cell.

The (E) shown in Eq. (5.52) is the emf needed to be applied at the process of electrolysis to initiate the electrochemical reaction. This voltage could be taken from a storage of electric or magnetic reservoir captured from induced current or voltage produced by lightning or switching surges.

It is also known that mass of substance accumulated during electrolysis or electroseparation is expressed by the equation:

$$Q \text{ is the charge density} = \frac{n\Gamma}{M} \text{ coulombs/kg} - \text{mole} \qquad (5.54)$$

where M = in equivalent kg-moles

Q_e = the electric charge in coulombs

Γ = Faraday's constant

n = the number of electrons released

Now, returning to Eq. (5.53), the quantity $E - E_0$ is the amount of potential difference needed to initiate the process of electrolysis and then possible electroseparation.

$$E - E_0 = \Delta E$$

$$\Delta E = \frac{RT}{n\Gamma} \ln \frac{P_C^c P_D^d}{P_A^a P_B^b} \text{ volts} \qquad (5.55)$$

Therefore,

$$n = \frac{RT}{\Gamma\Delta E} \ ln \ \frac{P_C^c P_D^d}{P_A^a P_B^b} \qquad (5.56)$$

and

$$Q_d = \frac{RT}{\Delta E} \ ln \ \frac{P_C^c P_D^d}{P_A^a P_B^b} \ \text{coulombs/kg} \qquad (5.57)$$

Total charge of electricity needed to transform H_2O to a mixture of H_2 and O_2 could be obtained by multiplying Q_d by the total mass of reactants released.

5.5.1 Shunt capacitance and series (source of electricity)

An example for a source of controlled flow of electricity is the system shown in Fig. 5.8. Voltage e_{t_2} and/or i_L is already stored by the capturing cycle due to lightning and/or switching surges (Fig. 5.8).

Time varying $e_{t_2}(t)$ is given by Eq. (4.93) and its steady-state value:

$$E_{t_2} = \frac{2A'Z_2L}{Z_1 + Z_2 + CA'Z_1Z_2} \qquad (5.58)$$

which, by replacing ΔE in Eq. (5.57), will give the required charge density in coulombs/kg.

Or, the time varying $i_L(t)$ given by Eq. (4.81) can be used as a current source that can be injected in the process of electrolysis and possible electroseparation for the ΔE indicated by Eq. (5.57).

$$\text{Average value of } i_L = \frac{2A'Z_2L}{Z_1 + Z_2 + CA'Z_1Z_2} \ \text{amperes} \qquad (5.59)$$

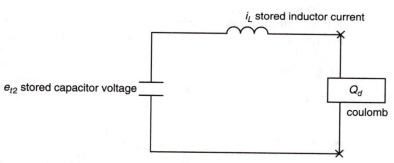

Figure 5.8 Generation of H_2-O_2 mixture by the release of Q_d.

This stored current source in the inductor can be transformed into a voltage source if it is preferred, or allowed to flow into the apparatus for electrolysis and then electroseparation.

$$i_L = \frac{dQ}{dt}$$

$$Q = \int_0^T i_L dt = \frac{2A'Z_2 LT}{Z_1 + Z_2 + CA'Z_1 Z_2} \text{ coulombs} \qquad (5.60)$$

where T is the total time of recharge in seconds.

If the weight of released reactants is equal to M kg, then from Eqs. (5.57) and (5.58), we can write:

$$\frac{MRT}{\Delta E} \ln \frac{P_A^a P_B^b}{P_C^c P_D^d} = \frac{2A'LT}{Z_1 + Z_2 + CA'Z_1 Z_2} \qquad (5.61)$$

$$M = \left(\frac{\Delta E}{RT}\right) \frac{2A'LT}{Z_1 + Z_2 + CA'Z_1 Z_2} \bigg| \ln \frac{P_A^a P_B^b}{P_C^c P_D^d} \text{ kg} \qquad (5.62)$$

Equation (5.62) is the total weight of reactants extracted from the electrochemical reaction cycle excited by current stored in the ideal inductor.

5.6 Exciting Current for the Induction Generator

As presented in Chap. 4, Sec. 4.7, magnetizing current for the three-phase induction generator has to be supplied from either an existing bus-bar or provided separately from a source of reactive power. Such a source could come from a storage of captured electromagnetic surges. A capacitor bank connected across the generator stator could be counted on to release controlled magnetizing current needed to produce rotating mmf in the air gap of the induction generator.

Equation (4.25) gives an expression for the induced voltage on a transmission line due to an ideal inducing lightning surge. Equation (4.29) shows the expression for the voltage charged across a high-quality capacitor having an initial stored voltage of V_0.

The magnetizing current is:

$$i_m = C \frac{dV_c(x,t)}{dt}$$

$$= C \frac{d}{dt} \left\{ V(x,t) + [V_0 - V(x,t)]e^{-\frac{t}{RC}} \right\} \qquad (5.63)$$

Since $V(x,t)$ is a step function, therefore,

$$i_m = CV(x,t) U_0(t)$$

$$+ \left[\frac{V(x,t)}{RC} U_{-1}(t) - V(x,t)U_0(t) - \frac{V_0}{RC} \right] e^{-\frac{t}{RC}} \tag{5.64}$$

If $V(x,t)$ in Eq. (4.29) comes from induced voltage at a transformer, it is given by Eq. (4.30) through (4.33), rewritten here for convenience:

$$V_{transformer}(x,t) = G(t)f(x) \tag{5.65}$$

where $f(x)$ is shown in Eq. (4.30) and

$$G(t) = A_2\delta(t) + A_3(A_c - A_v)[t - U_{-1}(t - t_1)] \tag{5.66}$$

Since

$$\frac{U_{-1}(t_1)}{c} \to 0$$

where c = velocity of light, therefore,
$$A_2 = E_0$$
$$A_3 = E_0/(A_c - A_v)$$
$$i_m = cf(x)\frac{d}{dt}G(t)$$
$$= cf(x)E_0\delta'(t) + E_0(A_c - A_v)[U_{-1}(t) - U_0(t - t_1)] \tag{5.67}$$

Magnetic field intensity vector $\overline{H}$ established in one phase of the generator stator is expressed by:

$$\overline{H} \cong \frac{N_{ph}i_m}{l} \quad \text{AT/meter} \tag{5.68}$$

where N_{ph} = number of stator turns per phase
   l = effective length of magnetic flux path

The magnetic induction vector $\overline{B}$ becomes:

$$\overline{B} = \mu \, \overline{H}$$

where μ = magnetic permeability of the stator core

Therefore, magnetic flux/phase Φ is given by:

$$\Phi(t) = \frac{\mu N_{ph}AE - o}{l} \{\delta'(t) + (A_c - A_v)[U_{-1}(t) - U_0(t - t_1)]\} \tag{5.69}$$

EMF generated/phase is shown as follows:

$$EMF(t) = N_{ph} \frac{d\Phi(t)}{dt}$$

$$= \frac{\mu N_{ph}^2 AE - 0}{l} \{\delta''(t)$$

$$+(A_c - A_v)[U_0(t) - U_0'(t - t_1)]\} \qquad (5.70)$$

To unify the singularities in Eq. (5.70):

$$EMF(t) = N_{ph} \frac{d\Phi(t)}{dt}$$

$$= \frac{\mu N_{ph}^2 AE - 0}{l} \{\delta''(t)$$

$$+ (A_c - A_v)[\delta(t) - \delta'(t - t_1)]\} \qquad (5.71)$$

because

$$U_0(t) = \delta(t)$$

5.7 Feasibility of MHD Induction Generator

Magnetic energy captured by a line inductor storage, in effect after considerable time the induced current produced by induction on a transmission line or at a transformer, could be used as the exiting source for a magnetohydrodynamic (MHD) induction generator. The line inductor could be geometrically wound around the MHD channel of rectangular cross section in close vicinity to the transmission line or transformer in an arrangement of $2N$ line conductors, N in the upstream channel and another N along the downstream channel. The induced current captured from lightning or switching surges will be allowed to flow across the rectangular MHD channel according to a predesigned control process.

A second set of three-phase line conductors will outwardly follow the inner set, in which a three-phase induced time-varying power will be developed, resulting from amplifying the larger number of turns of the second three-phase set of coils. This is shown in Fig. 5.9.

Each line conductor of the storing inductor will produce an x and a y component for the exciting applied magnetic field as follows:

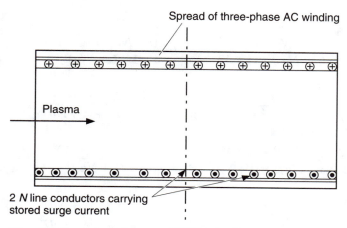

Spread of three-phase AC winding

Plasma

2 N line conductors carrying
stored surge current

Figure 5.9 Conceptual diagram of MHD induction generator.

$$B_{ox} = -i_{inductor} \frac{\mu_0}{2\pi} \frac{y + y_0}{[x^2 + (y + y_0)^2]} \qquad (5.72)$$

$$B_{oy} = i_{inductor} \frac{\mu_0}{2\pi} \frac{x}{[x^2 + (y + y_0)^2]} \qquad (5.73)$$

The separation among the set of the inductor turns is assumed equal to a unit length. The author indicated in previously published work that, by using the theory of mathematical perturbation in analyzing an MHD channel having conducting fluid of zero viscosity, subsonic, incompressible, and of moderate degree of ionization, the following solutions have been obtained for the induced magnetic field on a normalized nondimensional basis:

$$\hat{a}_x B = \frac{R_m}{2} \sum_{n=0}^{M} \frac{Y(x - N)}{(x - Na)^2 + (1 + Y)^2} + \sum_{n=0}^{\infty} \frac{L(x - N)}{(x - Na)^2 + (2Y + L_n)^2}$$

$$- \sum_{n=0}^{\infty} \frac{L(x - N)}{(x - Na)^2 + L_n^2} + \frac{R_m}{2} \sum_{n=0}^{M} \frac{Y(x + Na)}{(x - Na)^2 + (1 + Y)^2}$$

$$- \sum_{n=0}^{\infty} \frac{L(x - Na)}{(x - Na)^2 + (2Y + L_n)^2} - \sum_{n=0}^{\infty} \frac{L(x + Na)}{(x + Na)^2 + L_n^2} \qquad (5.74)$$

and

$$\hat{a}_y B = R_m \sum_{n=0}^{M} \frac{(x - Na)^2 + (HY)}{(x - Na)^2 + (1 + Y)^2} - \frac{1}{4} \ln \frac{(x - Na)^2 + (1 + Y)^2}{1 \div (x - Na)^2}$$

$$+ \frac{1}{4} \sum_{n=0}^{\infty} \frac{L_n + 2Y}{(x - Na)^2 + (L_n + 2Y)^2} + \frac{L}{2} \sum_{n=0}^{\infty} \frac{L_n^2 - (x - Na)^2}{L_n^2 + (x + Na)^2} Y$$

$$+ \frac{R_m}{2} \frac{(x + Na)^2 + (1 + Y)}{(x - Na)^2 + (1 + Y)^2} + \frac{1}{4} \ln \frac{(x + Na)^2 + (1 + Y)^2}{1 + (x + Na)^2}$$

$$+ \frac{1}{4} \sum_{n=0}^{\infty} \frac{Ln + 2Y}{(x + Na)^2 + (L_n + 2Y)^2} + \frac{L}{2} \sum_{n=0}^{\infty} \frac{L_n^2 - (x + Na)^2}{L_n^2 + (x + Na)^2}$$

$$+ \frac{1}{2} \tag{5.75}$$

where L = channel length or inductor length
$L_n = 1 + L + 2nL$
$2N$ = total number of turns for the storing inductor or, in effect, the number of line conductors across the MHD induction generator
R_m = magnetic Reynold number, physically known as the ratio of the induced to the applied magnetic induction $\bar{B}$

Moderate degree of ionization for the conducting fluid or plasm translates into $0.01 \leq R_m \leq 0.1$.

Design of the MHD induction generator in this case presumes that arrangement of the line conductors across the MHD channel be in the form of alternating apposite polarities where the y component of the inducted $\bar{B}$ field will be almost neutralized, as shown in Fig. 5.10, while the x component of $\bar{B}$ is expanded using the Fourier theorem:

$$\hat{a}_x B = \sum_{n=1}^{\infty} \frac{9A}{n^2 \pi^2} \sin \frac{n\pi}{3} \sin \frac{n\pi x}{a} - \sum_{n=1}^{\infty} \frac{6A}{n\pi} \sin \frac{n\pi x}{a} \tag{5.76}$$

where A is the peak of $\hat{a}_x \bar{B}$ vector which could be a function of x and t.
Now the total flux linkage Φ is as follows:

$$\Phi = \int_{-l}^{l} \int_{a_i}^{a_0} Bx \tag{5.77}$$

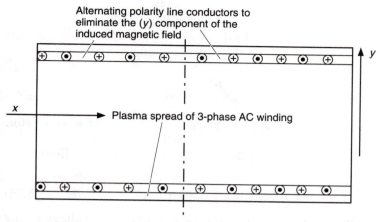

Figure 5.10 Conceptual design of linear MHD induction generator.

Then, assuming an ideal case where all the turns per phase are linked simultaneously, i.e., all turns are confined to one equivalent slot, therefore, from the fact:

$$aB < \frac{MMF}{\text{magnetic reluctance of air path}} \qquad (5.78)$$

where MMF = magnetomotive force
  a = area of flux linkage with $\mu = 1$ (magnetic permeability)

Therefore,

$$B < \frac{MMF}{\text{length of magnetic path}} \qquad (5.79)$$

Therefore,

$$MMF = B(x)X \qquad (5.80)$$

where X is a variable channel length along the reactor divertor.

Compiling Eqs. (5.77), (5.78), (5.79), and (5.80) and then seeking a condition for maximum fundamental MMF:

$$MMF = \text{A constant} \qquad (5.81)$$

or that

$$(x)B(x) = \text{A constant} = K \qquad (5.82)$$

Therefore,

$$B(x) = \frac{K}{X}$$
(5.83)

Selecting:

$$X = Ca$$
(5.84)

where C = number of conductors placed at one side of the divertor, either in the upstream region or the downstream region

It has been determined that for peak MMF the optimum limit for X along the divertor channel is:

$$X = n\pi \; (n \text{ is an odd integer})$$
(5.85)

It can be concluded that for complete harmonic neutralization (by forming a three-phase connection), maximum value for a fundamental MMF can be secured if one-half of the divertor channel is an odd multiple of separation among the group of the linear superconductors.

Calculations of the induced magnetic fields revealed the fact that both the tangential and transverse components are zero at or even near the divertor surface. Also, it is found that B_x and B_y are within the tube of the diverted plasma.

By adding another set of idle coils with a relatively large number of turns and feeding a time-varying sinusoidal current in the first or internal divertor coil, an AC power output could be obtained from the secondary coil by the induction principle. This amounts to the performance of a linear induction type of MHD generator.

Maximum effective *MMF* for voltage induction in the secondary coil may be obtained if either coil (the exciting coil or the secondary coil) is confined to a short axial length along the divertor channel.

While it is feasible that a linear induction MHD generator could be designed using a slight modification in the divertor channel, an extension of the channel to a cylindrical structure for the plasma flow may make probable the design of an MHD synchronous generator as a topping machine for electric AC power output.

5.8   Feasibility of Electrogasdynamic (EGD) Generator

Voltage charged across a condenser installed to capture induced voltage on a transformer or transmission line could be used to operate an electrogasdynamic converter. Application of the captured electric potential by the storing capacitance across a channel of conducting fluid with positive ions domination is to be performed through a process of controlled release of the stored electric energy.

Voltage $V_c(x,t)$ captured by the storing capacitor at which V_0 was existing is given by Eq. (4.29), rewritten here for convenience:

$$V_c(x,t) = V(x,t) + [V_0 - V(x,t)]e^{-\frac{t}{RC}} \qquad (5.86)$$

$V(x,t)$ is the induced electric potential difference on a transmission line which is expressed by Eq. (4.25) or on a transformer which is given by Eq. (4.33), both produced by lightning surge or switching surges.

Assuming uniformity of the electric field axially along the EGD generator, the axial electric field $E_{axial} = E_x = V_c(x,t)/l$, where l is the channel length:

$$E_x = \frac{V_c(x,t)}{l} \text{ volts/meter} \qquad (5.87)$$

Let E_r be the radial electric field caused by space charge effect. Therefore, the equation of balance between the axial and radial fields prior to breakdown is given by:

$$E_x = E_r \qquad (5.88)$$

where in one dimension, the space charge density presence is given by the equation:

$$\frac{\delta E_x}{\delta x} = \rho_e/\epsilon \qquad (5.89)$$

where e is the dielectric permitivity of EGD channel.

For v, the mobility of positive ions, the drift velocity is given by vE_x, and, therefore, the net moving velocity $v_m = v_g - v_d$, where v_g is the gas-free velocity.

The generator current density J_x is given by:

$$J_x = -\rho v_m \qquad (5.90)$$

And, hence, the power output/unit area of electrode:

$$P_A = J_x V_c(x,t) \qquad (5.91)$$

5.9 Problems

5.1 Develop a mathematical model for the storage battery in the complex frequency domain, representing the circuit model of that shown in Fig. 5.1.

5.2 Develop an expression for the weight of material deposition in a process of recharge of an ideal storage battery when controlled electric current is allowed to flow into the battery from energy captured by the inductor system installed at a location of transmission line.

5.3 Repeat Prob. 5.2 when a voltage is allowed to be applied in a controlled process from energy captured by an ideal capacitor installed at a location on a transmission line.

5.4 Repeat Prob. 5.2 if the captured energy is coming from an ideal inductor installed at the entry of transformer.

5.5 Repeat Prob. 5.3 if the captured energy is coming from an ideal capacitor installed at the entry of transformer.

5.6 Repeat Prob. 5.2 if the flow of current into the battery is coming from either an inductor connected in parallel with a capacitor or from the voltage stored by that capacitor itself.

5.7 Determine a criterion for the required charge density required in coulomb/kg to recharge a $Ti - Fe$ redox flow cell when it is subjected to controlled voltage across a series capacitor located on a transmission line in shunt with an inductor.

5.8 Repeat Prob. 5.7 if the series capacitor and the shunt inductor are installed at the entry of transformer.

5.9 Repeat Prob. 5.7 if the system of L-C is in the form of series inductor and shunt capacitance.

5.10 Repeat Prob. 5.8 if the system of L-C is in the form of series inductor and shunt capacitance.

5.11 In the process of electrolysis and electroseparation of $H_2 - O_2$, obtain the solution of total weight of reactants (i.e., hydrogen and oxygen) if DE required in Eq. (5.41) is secured from the voltage stored in the capacitance shown in Fig. 5.8.

5.12 Repeat the problem if ΔE is extracted from an ideal single capacitance inserted in series on a transmission line, and having captured voltage induced due to lightning surge. Discuss what will happen at resonance.

5.13 Repeat Prob. 5.12 if the captured voltage induced by the capacitor came from the entry of the transformer subjected to lightning surge.

5.14 Repeat Prob. 5.7 if the L-C system is in the form of series capacitance and shunt inductor.

5.15 Repeat Prob. 5.7 if the L-C system is in the form of a parallel combination inserted in series in the transmission line. Discuss what will happen at resonance.

5.16 Consider the electromechanical induction generator is provided with its controlled magnetizing current from an ideal inductor containing captured magnetic energy from induced lightning surge on a transmission line. Obtain the expression for the produced flux/phase similar to Eq. (5.67). Also, give an expression for the EMF developed similar to Eq. (5.71).

5.17 Repeat Prob. 5.16 if the stored current in the inductor has been captured from the induction effect at the entry of the transformer.

5.18 For an MHD induction generator excited by controlled current from an inductor having captured magnetic energy, modify Eq. (5.57) if the captured energy by the inductor is from the effect of induction at the entry of the transformer to which the inductor is connected.

5.19 For an EGD generator, establish a solution for the power density developed if its axial electric field is produced by a controlled process from a series capacitance across an inductor combination, and the whole thing is in series through a transmission line that was subjected to an ideal lightning surge.

5.10 References

1. Bockris, J. O'M., and Reddy, A.K.N., *Modern Electrochemistry*, vol. 142, Plenum Press, New York, 1971.
2. Denno, K., "Auxiliary Control of the Magnetic Field System of the Fusion Reactor Divertor," *Proceedings of the Sixth Symposium on Engineering Problems of Fusion Research*, San Diego, Calif., Nov. 1975, pp. 784–787.
3. Denno, K., and Fouad, A. A., "Effects of the Induced Magnetic Fields on the MHD Channel Flow," *IEEE Transactions on Electron Devices*, vol. 19, March 1972.
4. Denno, K., "Behavior of Weakened Fusion Plasma in MHD Channel," a paper submitted for possible publication.
5. Denno, K, "Generation Aspects of Weakened Fusion Plasma in MHD channel," a paper submitted for possible publication.
6. Denno, K., "Mathematical Modelling of Storage Battery and Fuel-Cell," *Proceedings IEEE International Telephone Energy Conference*, Washington, D.C., 1978, pp. 237–243.
7. Denno, K., *Power System Design and Applications for Alternative Energy Sources*, Prentice-Hall, Inc., Englewood Cliffs, N.J. 1989.
8. Denno, K., *Engineering Economics of Alternative Energy Sources*, CRC Press, Boca-Raton Florida, 1990.
9. Denno, K., *High Voltage Engineering in Power Systems*, CRC Press, Boca Raton, Florida.
10. File, J., "Proposed Superconducting Coils for the Princeton Fusion Reactor," *Proceedings of the Fifth Symposium on Engineering Problems of Fusion Research*.
11. Fish, J. D., and Axtmann, R. C., "Utilization of Plasma Exhaust Energy for Fuel Production," *Proceedings of the Fifth Symposium on Engineering Problems of Fusion Research*.
12. Grainick, S., "The operating cycle of a Long Pulse Tokamak Fusion Reactor," *Proceedings of the Fifth Symposium on Engineering Problems of Fusion Research*.
13. Soo, S. L., *Direct Energy Conversion*, Prentice-Hall, Inc., Englewood Cliffs, N.J., 1968.
14. Tenney, F. H., "A 2000 MW Fusion Reactor, an overview," *Proceedings of the Fifth Symposium on Engineering Problems of Fusion Research*.
15. Walsh, E. M., *Energy Conversion*, The Ronald Press Co., New York, 1967.

Protective Devices Comprising Saturistor with Either Ideal Inductor or Capacitor

In Chap. 4, the author presented mechanisms of protective devices containing ideal inductor, ideal capacitor, and a combination of inductors and capacitors in various series-parallel forms. Solutions for output voltage across an ideal capacitor, or current stored by the ideal inductor, have been obtained in each network configuration. In reference to induced voltage or current resulting from lightning or switching surges, expressions for energy stored were established. Induced voltages on a transmission line or at the entry of a transformer have been considered as the source for transferring that voltage to an ideal capacitor and the current surge to the ideal inductor. The action performed by the ideal capacitor or ideal inductor for capturing the induced effects of lightning and switching surges was, in effect, a practical function of protection. Of course, the capturing performance by L or C has to be initiated by a special switching-on-or-off mechanism with respect to time and position.

In Chap. 5, voltage and current captured by the ideal capacitor or inductor, and by combination of L and C in various series-parallel connections, have been systemized for direct or indirect utilization by electrochemical, electromechanical, and direct energy converters to produce feasible electric power output. Electrochemical power generators include the storage battery (also known as *accumulators*), the redox flow cell which operates on the basis of oxidation = reduction, the process of electrolysis and electroseparation for the production of H_2 and O_2, the electromechanical induction generator to which the initial magnetizing current is injected from current captured by an ideal inductor, and the magnetohydrodynamic generator to which the exciting current for its working fluid could come from either an inductor or capacitor.

In Chap. 2, comprehensive analysis for the steady-state and transient characteristics of the saturable-resistor were presented. It was stated that the saturistor (saturable-resistor) is a reactor whose resistance, reactance, and impedance characterizes a special functional behavior with respect to a passing time-varying current, and also demonstrates a linear function with respect to frequency.

In this chapter, the protective role of the saturable-resistor in a scheme of either series connection or parallel connection with the ideal capacitor and with the ideal inductor will be analyzed. We are going to secure general solutions for the terminal voltage of each device to indicate clearly the protective function in reducing or weakening the incident voltage or current surges.

Protective devices that will be treated in this chapter include: (1) series combination of saturistor with an ideal inductor or ideal capacitor and (2) a series saturistor with a shunt ideal capacitor and parallel scheme involving an ideal inductor or capacitor with the equivalent resistance or equivalent inductance of the saturable-resistor.

In the presentation to follow, the functional form for the ohmic values of the saturable-resistor is considered a straight-line representation, as shown in Fig. 6.1 and expressed as follows:

$$L_s - before = \frac{L_s - max}{I - i_p} i(t) + \frac{i_p L_{s - max}}{i_p - I} \tag{6.1}$$

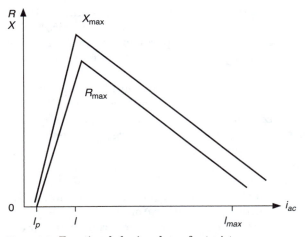

Figure 6.1 Functional ohmic values of saturistor.

and

$$R_{s-before} = \frac{R_{s-max}}{I - i_p} \, i(t) + \frac{i_p}{i_p - I} \, R_{s-max} \tag{6.2}$$

Equations (6.1) and (6.2) are valid for $i_p < i < I$.

The second term of Eq. (6.2) is, in effect, the damping resistive element of the saturable-resistor. Then,

L_s, R_s is the saturable-resistor inductance and resistance at any $i(t)$.

L_{s-max}, R_{s-max} is the maximum inductance and resistance of the saturable-resistor which occurs at $i(t) = I$.

i_p is the value of the saturable-resistor current at which its resistance and inductance starts to pick up, respectively.

It is a clear point to keep in mind that the saturable-resistor provides the function for protection when $i(t) = I$, where I is the current flow in the saturable-resistor at which its ohmic values are maximum, thereby providing the most desirable protection in all cases. Also, since saturable-resistor has a constant resistance before the pickup level, we are in effect including the resistor damper in the process of protection provided by the saturable-resistor.

6.1 Saturable-Resistor in Series with Ideal Inductor

This is shown in Fig. 6.2.

From the principles of circuit theory, namely, basic Kirchoff's laws, we can write the following equations of continuity for voltages and currents:

$$e_{t_1} + e_{r_1} = L \frac{di_1}{dt} + L_s \frac{di_1}{dt} + R_s i_1 + e_{t_2} \tag{6.3}$$

$$i_1 = i_2 \tag{6.4}$$

Saturable-resistor

Figure 6.2 Ideal inductor and saturistor.

or
$$e_{t_1} + e_{r_1} = \frac{L + L_s}{Z_1} \frac{de_{t_1}}{dt} + R_s \frac{e_{t_1}}{Z_1} + e_{t_2} \qquad (6.5)$$

$$e_{t_1} - e_{r_1} = \frac{Z_1}{Z_2} e_{t_2} \qquad (6.6)$$

where Z_1, Z_2 is the surge impedance of the incoming line and outgoing line, respectively.

e_{t_1} is the transmitted surge into line 1.

e_{t_2} is the transmitted surge into line 2.

e_{r_1} is the reflected surge back into line 1.

Now, adding Eqs. (6.5) and (6.6) gives:

$$2e_{t_1} = \frac{Z_1}{Z_2} e_{t_2} + e_{t_2}$$

$$+ \frac{1}{Z_1} \left[L + L_0 + \frac{L_{s-max}}{I - i_p} \frac{e_{t_1}}{Z_1} \right] \frac{de_{t_1}}{dt}$$

$$+ \frac{1}{Z_1} \left[R_0 + \frac{L_{s-max}}{I - i_p} \frac{e_{t_1}}{Z_1} \right] e_{t_1} \qquad (6.7)$$

where e_{t_1} is the transmitted or, in effect, the induced voltage on a transmission line or at the entry of the transformer or tower. The induced voltage on a transmission line due to an ideal lightning surge was given by Eq. (4.25), while that on transformer was given by Eq. (4.30).

Equation (4.25) specifies the surge voltage induced due to an ideal sustained lightning surge having the functional form of $AU_{-1}(t)G(x)$, while that induced on a transformer is of the form $A(t)G(x)$.

Referring back to Eq. (6.7), R_0 and L_0 is the saturable-resistor's resistance and inductance, respectively, up to when $i(t) = i_p$ (the pickup current), after which all ohmic values start to rise.

Taking the case for the voltage induced on a transmission line due to a sustained lightning surge, e_{t_2} induced at the terminal of the scheme shown in Fig. 6.7 is given as:

Where

$$e_{t_1}(t) = AU_{-1}(t)G(x)$$

$$e_{t_2} = \frac{Z_2}{Z_1 + Z_2} \left\{ 2AU_{-1}(t) \right.$$

$$-\frac{1}{Z_1}\left[L+L_0+\frac{L_{s-max}}{I-i_p}\frac{A}{Z_1}U_{-1}(t)\right]A\delta(t)$$

$$-\frac{1}{Z_1}\left[R_0+\frac{L_{s-max}}{I-i_p}\frac{AU_{-1}(t)}{Z_1}\right]AU_{-1}(t)\Bigg\}G(x) \qquad (6.8)$$

e_{t_2}, given by Eq. (6.8) is valid for the range of $i(t)$, larger than i_p but less than I, which is the current flow into the saturable-resistor for maximum resistance and inductance.

Another scope of change for i(t) is:

$$i < i(t) < i_{saturation} \qquad (6.9)$$

where $I_{saturation}$ is the maximum value of $i(t)$ when both the saturable-resistor inductance and resistance become ultimately zero (which from now on will be referred to as I_{max}). $L_{s-after}$ and $R_{s-after}$, corresponding to the inequality of Eq. (6.9), are given as:

$$L_{s-after} = \frac{L_{s-max}}{I-I_{max}}\,i(t) + \frac{I_{max}L_{s-max}}{I_{max}-I} \qquad (6.10)$$

and

$$R_{s-after} = \frac{R_{s-max}}{I-I_{max}}\,i(t) + \frac{I_{max}R_{s-max}}{I_{max}-I} \qquad (6.11)$$

The second term of Eq. (6.11) is, in effect, the damping resistive element.

Using Eqs. (6.5) and (6.6) with Eqs. (6.10) and (6.11), the following differential equation is obtained:

$$e_{t_2} = \left(\frac{Z_2}{Z_1+Z_2}\right)\Bigg\{2e_{t_1} - \frac{1}{Z_1}\left[L+\frac{e_{t_1}}{Z_1}\frac{L_{s-max}}{I-I_{max}}+\frac{I_{max}}{I_{max}-I}\right]\frac{de_{t_1}}{dt}$$

$$+\frac{e_{t_1}}{Z_1}\left[\frac{R_{s-max}}{I-I_{max}}\frac{e_{t_1}}{Z_1}+\frac{I_{max}}{I_{max}-I}R_{s-max}\right]\Bigg\} \qquad (6.12)$$

The solution for e_{t_2} at the terminals of the protective series combination of ideal inductor and saturable-resistor is given as follows:

$$e_{t_2}(t) = \frac{AZ_2G(x)}{Z_1-Z_2}\Bigg\{2U_{-1}(t) - \frac{\delta(t)}{Z_1}\left[L+\frac{AU_{-1}(t)}{Z_1}\frac{L_{s-max}}{I-I_{max}}\frac{I_{max}}{I_{max}-I}+\frac{I_{max}}{I_{max}-I}\right]$$

$$+\frac{U_{-1}(t)}{Z_1}\left[\frac{AU_{-1}(t)}{Z_1}\frac{R_{s-max}}{I-I_{max}}+\frac{I_{max}}{I_{max}-I}R_{s-max}\right]\Bigg\} \qquad (6.13)$$

Analysis of e_{t_2} when $i(t) = I = A/Z_1$ will follow:

Z_1 is the surge impedance of the incoming line.

A is the magnitude of the incident induced surge $e_{t_1}(t)$.

At $t > 0$ and $i(t) = 1$, then from Eq. (6.8), e_{t_2} is given by:

$$e_{t_2} = \frac{Z_2 G(x)I}{Z_1 + Z_2}\left[2Z_1 - \left(R_0 + \frac{I}{I - i_p} R_{s-max}\right)\right]U_{-1}(t) \qquad (6.14)$$

Equation (6.14) clearly pinpoints the peak role of the saturable-resistor resistance in weakening the transmitted surge into line 2.

Then for $i(t) = I_{max}$ and $t > 0$.

$$e_{t_2} = \frac{IZ_2 G(x)}{Z_1 + Z_2}\left\{2Z_1 + \left(I \frac{R_{s-max}}{I - I_{max}} + R_{s-max} \frac{I_{max}}{I_{max} - I}\right)\right\} \qquad (6.15)$$

e_{t_2} also shows the minimal effect of the saturable-resistor in weakening e_{t_2} since $(I - I_{max})$ is negative.

6.2 Saturable-Resistor in Series with Ideal Capacitor

Refer to Fig. 6.3. Continuity of circuit theory equations at lines 1 and 2 gives:

$$e_{t_1} + e_{r_1} = e_{t_2} + L_s \frac{di_{t_1}}{dt} + \frac{1}{c}\int i_{t_1}dt + R_s i_{t_1} \qquad (6.16)$$

$$i_{t_1} + i_{r_1} = i_{t_2} \qquad (6.17)$$

$$i_{t_1} = \frac{e_{t_1}}{Z_1} \qquad (6.18)$$

Figure 6.3 Series C, L, and saturistor.

Therefore,

$$\frac{e_{t_1}}{Z_1} - \frac{e_{r_1}}{Z_1} = \frac{e_{t_2}}{Z_2} \tag{6.19}$$

From Eqs. (6.16) and (6.19),

$$2e_{t_1} = \frac{Z_1}{Z_2} e_{t_2} + e_{t_2} + \frac{L_s}{Z_1} \frac{de_{t_1}}{dt} + \frac{1}{cZ_1} \int e_{t_1} dt + R_s i_{t_1} \tag{6.20}$$

Voltage induced on a transmission line from a sustained lightning surge identifies $e_{t_1} = AU_{-1}(t)G(x)$.

From Eq. (6.20) with $e_{t_1} = AU_{-1}(t)G(x)$ and when current flow in the saturable-resistor i_{t_1} is less than I but larger than the pickup current i_p, then by using Eqs. (6.1) and (6.2) in describing $L_{s-before}$ and $R_{s-before}$, the solution for e_{t_2}, the transmitted voltage surge into line 2, whose surge impedance Z_2, is given by:

$$e_{t_2} = \frac{Z_2 G(x)}{Z_1 + Z_2} \left\{ 2AU_{-1}(t) - \frac{A\delta(t)}{Z_1} \left[\frac{L_{s-max}}{I - i_p} \frac{AU_{-1}(t)}{Z_1} + \frac{i_p}{i_p - I} L_{s-max} \right] \right.$$

$$\left. - \frac{AU_{-1}(t)}{Z_1} \left[\frac{R_{s-max}}{I - i_p} \frac{AU_{-1}(t)}{Z_1} + \frac{i_p}{i_p - I} R_{s-max} \right] - \frac{At}{cZ_1} \right\} \tag{6.21}$$

At $t > 0$, the term preceded by $\delta(t)$ becomes zero, e_{t_2} is given by:

$$e_{t_2} = \frac{AZ_2 G(x)}{Z_1 + Z_2} \left\{ 2Z_1 - \left[\frac{R_{s-max}}{I - i_p} AU_{-1}(t) - \frac{i_p}{i_p - I} R_{s-max} \right] \right.$$

$$\left. - \frac{AU_{-1}(t)Z_2}{c(Z_1 + Z_2)} t \right\} \tag{6.22}$$

We remind the reader that Eq. (6.22) is valid when current passing into the saturable-resistor is beyond the pickup current i_p but less than the current for maximum ohmic values. In this range the saturable-resistor provides a rising protective role which will peak when $i_{t_1}(t) = I$.

Next, considering the second phase regarding the situation when current flow in the saturable-resistor is beyond I but less than the value when all the ohmic values of the saturable-resistor become zero, such a value being identified as I_{max}, the solution for the transmitted voltage surge into line 2 is given by:

$$e_{t_2} = \frac{Z_2 G(x)}{Z_1 + Z_2} \left\{ 2AU_{-1}(t) - \frac{A\delta(t)}{Z_1} \left[\frac{L_{s-max}}{I - I_{max}} \frac{AU_{-1}(t)}{Z_1} + \frac{I_{max}}{I_{max} - I} L_{s-max} \right] \right.$$

$$\left. - \frac{AU_{-1}(t)}{Z_1} \left[\frac{R_{s-max}}{I - I_{max}} \frac{AU_{-1}(t)}{Z_1} + \frac{I_{max}}{I_{max} - I} R_{s-max} \right] - \frac{At}{cZ_1} \right\} \tag{6.23}$$

And, at $t > 0$, e_{t_2} becomes:

$$e_{t_2} = \frac{Z_2 G(x)}{Z_1 + Z_2} \left\{ 2AU_{-1}(t) - \frac{AU_{-1}(t)}{Z_1} \left[\frac{R_{s-max}}{I - I_{max}} \frac{AU_{-1}(t)}{Z_1} \right. \right.$$

$$\left. \left. + \frac{I_{max}}{I_{max} - I} R_{s-max} \right] - \frac{At}{cZ_1} \right\} \qquad (6.24)$$

Inspection of Eqs. (6.22) and (6.24) demonstrate clearly the effective but declining role of the resistive component of the saturable-resistor in weakening the stressful level of the transmitted voltage surge into line 2.

6.3   Shunt Inductor and a Series Saturistor

This scheme is shown in Fig. 6.4.

Equations of continuity for currents and voltages for the system shown in Fig. 6.4 are given as follows:

$$e_{t_1} + e_{r_1} = e_{t_2} + R_s i_{t_1} + L_s \frac{di_{t_1}}{dt} \qquad (6.25)$$

$$i_{t_1} = \frac{e_{t_1}}{Z_1} \qquad (6.26)$$

or

$$i_{t_1} + i_{r_1} = \frac{1}{L} \int e_{t_1} dt + i_{t_2} \qquad (6.27)$$

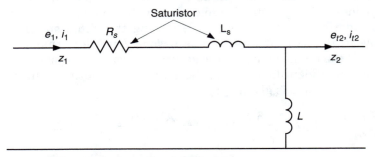

Figure 6.4 Shunt inductor and series saturistor.

Therefore,

$$\frac{e_{t_1}}{Z_1} - \frac{e_{r_1}}{Z_1} = \frac{1}{L} \int e_{t_2} dt + \frac{e_{t_2}}{Z_2} \tag{6.28}$$

$$e_{t_1} - e_{r_1} = \frac{Z_1}{L} \int e_{t_2} dt + \frac{Z_1}{Z_2} e_{t_2} \tag{6.29}$$

From Eqs. (6.25) and (6.29), we write:

$$2e_{t_1} = e_{t_2} + R_s \frac{e_{t_1}}{Z_1} + \frac{L_s}{Z_1} \frac{de_{t_1}}{dt} + \frac{Z_1}{L} \int e_{t_2} dt + \frac{Z_1}{Z_2} e_{t_2} \tag{6.30}$$

Therefore,

$$e_{t_2} \left(1 + \frac{Z_1}{Z_2}\right) + \frac{Z_1}{L} \int e_{t_2} dt = 2e_{t_1} - \frac{R_s}{Z_1} e_{t_1} - \frac{L_s}{Z_1} \frac{e_{t_1}}{dt} \tag{6.31}$$

The saturable-resistor components of $L_{s - before}$ and $R_{s - before}$ are expressed by Eqs. (6.1) and (6.2), which could be rewritten as follows:

$$L_{s - before} = a \frac{e_{t_1}}{Z_2} + b \tag{6.32}$$

and

$$R_{s - before} = c \frac{e_{t_1}}{Z_2} + f \tag{6.33}$$

Equations (6.32) and (6.33) are valid for $i_p < i(t) < I$.

Substitution for the mathematical forms of $L_{s - before}$ and $R_{s - before}$, which are a function of e_{t_1}, into Eq. (6.31) will result in having terms with the forms $(e_{t_1})^2$ and $[e_{t_1} de_{t_1}/dt]$. To obtain the solution for e_{t_2} using Laplace transforms, the following special rule has been used to secure the general solution for the transmitted voltage surge into line 2, namely, e_{t_2}:

$$\mathcal{L} f_1(t) f_2(t) = \sum_{p=1}^{k} \frac{A_1(s_k)}{B_1'(s_k)} F_2(s - s_k) \tag{6.34}$$

where

$$\mathcal{L} f_1(t) = \frac{A_1(s)}{B_1(s)} \quad \text{a rational function} \tag{6.35}$$

$B_1(s)$ has k first-order poles.

Since $e_{t_1}(t) = AU_{-1}(t)G(x)$ as the induced voltage on a transmission line due to a sustained lightning surge, and using the preceding rule of Eq. (6.34), the following transforms have been secured:

$$\mathcal{L}AU_{-1}(t) = \frac{A}{S} = \mathcal{L}f_1(t) \tag{6.36}$$

$$\mathcal{L}e'_{t_1} = S\frac{A}{S} = A \quad \text{with zero initial condition} \tag{6.37}$$

$$\mathcal{L}e_{t_1}(t)e'_{t_2}(t) = \frac{A}{1} = A^2 \tag{6.38}$$

$$\mathcal{L}e_{t_1}(t)e_{t_2}(t) = \frac{A}{1}\frac{A}{S} = \frac{A^2}{S} \tag{6.39}$$

Applying the process of Laplace transform on Eq. (6.31) involving the special rule of Eqs. (6.34) and (6.35), the solution for e_{t_2} is given following for $i_p < i(t) < I$, where $i(t)$ is the current flow in the saturable-resistor (i.e., i_{t_1} in this case):

$$e_{t_2}(t) = \frac{K_1Z_2}{Z_1+Z_2} e^{-\frac{Z_1Z_2}{L(1+Z_2)}t} - \frac{K_2Z_2}{(1+Z_2)}\delta(t) - \frac{LK_2}{Z_1}U_{-1}(t) \tag{6.40}$$

where

$$K_1 = \frac{A}{Z_1}\left[(2Z_1-f) - \frac{CA}{Z_1}\right]G(x) \tag{6.41}$$

$$K_2 = \frac{A}{Z_1}\left[\frac{a'A}{Z_1} + Ab\right]G(x) \tag{6.42}$$

Equations (6.1) and (6.2) give a, b, c, and f, rewritten below for reader convenience:

$$a = \frac{L_{s-max}}{I-i_p}$$

$$b = \frac{i_pL_{s-max}}{i_p-I}$$

$$c = \frac{R_{s-max}}{I-i_p}$$

$$d = \frac{i_pR_{s-max}}{i_p-I}$$

At $t > 0$,

$$e_{t_2} = \frac{K_1 Z_2}{Z_1 + Z_2} e^{-\frac{Z_1 Z_2}{L(1+Z_2)}t} - \frac{LK_2}{Z_1} U_{-1} \tag{6.43}$$

where $e_{t_2}(t)$ of Eq. (6.40) points to a requirement at which the saturable-resistor provides its rising protective function up to the point where $i_{t_1}(t) = I$.

Treating the case when surge current flow into the saturable-resistor $i(t)$ is larger than I but less than I_{max} (current when the ohmic components of the saturable-resistor are at minimum), $L_{s-after}$ and $R_{s-after}$ that were expressed in Eqs. (6.10) and (6.11) are given in a simpler form as follows:

$$L_{s-after} = a'i(t) + b' \tag{6.44}$$

$$R_{s-after} = c'i(t) + f' \tag{6.45}$$

where

$$a' = \frac{L_{s-max}}{I - I_{max}} \tag{6.46}$$

$$b' = \frac{I_{max} L_{s-max}}{I_{max} - I} \tag{6.47}$$

$$c' = \frac{R_{s-max}}{I - I_{max}} \tag{6.48}$$

$$f' = \frac{I_{max} R_{s-max}}{I_{max} - I} \tag{6.49}$$

Returning to the differential equation of Eq. (6.31) to solve for $e_{t_2}(t)$ when the surge in the saturable-resistor is beyond the level corresponding to maximum ohmic values, the following is obtained for $e_{t_2}(t)$:

$$e_{t_2}(t) = \frac{K_1' Z_2}{Z_1 + Z_2} e^{-\frac{Z_1 Z_2}{L(1+Z_2)}t} - \frac{K_2' Z_2}{1 + Z_1} \delta(t) - \frac{LK_2'}{Z_1} U_{-1}(t) \tag{6.50}$$

And, of course, for $t > 0$:

$$e_{t_2}(t) = \frac{K_1' Z_2}{Z_1 + Z_2} e^{-\frac{Z_1 Z_2}{L(1+Z_2)}t} - \frac{LK_2'}{Z_1} U_{-1}(t) \tag{6.51}$$

where

$$k_1' = \frac{A}{Z_1} \left[(2Z_1 - f') - \frac{c'A}{Z_1} \right] G(x) \tag{6.52}$$

$$k'_2 = \frac{A}{Z_1}\left[\frac{a'A}{Z_1} + Ab'\right]G(x) \tag{6.53}$$

Let us examine k'_1 and k'_2:

$$k'_1 = \frac{A}{Z_1}\left[2Z_1 - \frac{I_{max}R_{s-max}}{I_{max}-I} - \frac{A}{Z_1}\frac{R_{s-max}}{I-I_{max}}\right]G(x) \tag{6.54}$$

The indications in Eq. (6.54) are that the second term is positive while the third term is negative, but the high probability is that the second term is larger than the first term, rendering the net effect for k'_1 a weakening factor due to the resistive component of the saturable-resistor. We have to remember that A is the magnitude of the incident or induced voltage surge on the transmission line and, therefore, A/Z_1 is the surge-induced current, or i_{t_1}.

Turning next to the factor k'_2:

$$k'_2 = \frac{A}{Z_1}\left[\frac{A}{Z_1}\frac{L_{s-max}}{I-I_{max}} - A\frac{I_{max}L_{s-max}}{I_{max}-I}\right]G(x) \tag{6.55}$$

It is very obvious that the second term in Eq. (6.55) is positive and larger than the first term, which is negative, rendering a net weakening effect on e_{t_2} described in Eq. (6.50) due to the inductive effect of the saturable-resistor; however, in this region the protective role is declining.

6.4 Series Saturable-Resistor and a Shunt Ideal Capacitor

This is shown in Fig. 6.5.

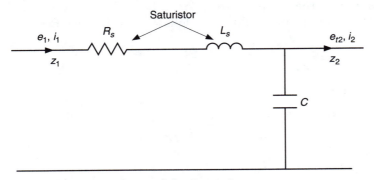

Figure 6.5 Series saturistor and shunt capacitance.

Equations for circuit theory continuity, which are based on Kirchoff's laws, are:

$$e_{t_1} + e_{r_1} + L_s \frac{di_{t_1}}{dt} + R_s i_{t_1} = e_{t_2} \tag{6.56}$$

or

$$e_{t_1} + e_{r_1} + \frac{L_s}{Z_1} \frac{de_{t_1}}{dt} + \frac{R_s}{Z_1} e_{t_1} = e_{t_2} \tag{6.57}$$

and

$$i_{t_1} + i_{r_1} = C \frac{de_{t_2}}{dt} + i_{t_2} \tag{6.58}$$

or

$$\frac{e_{t_1}}{Z_1} - \frac{e_{r_1}}{Z_1} = C \frac{de_{t_2}}{dt} + \frac{e_{t_2}}{Z_2} \tag{6.59}$$

Therefore,

$$e_{t_1} - e_{r_1} = C Z_1 \frac{de_{t_2}}{dt} + \frac{Z_1}{Z_2} e_{t_2} \tag{6.60}$$

Adding Eqs. (6.56) and (6.59) gives:

$$2e_{t_1} = C Z_1 \frac{de_{t_2}}{dt} + \frac{Z_1}{Z_2} e_{t_2} - e_{t_2} - \frac{L_s}{Z_1} \frac{de_{t_1}}{dt} - \frac{R_s}{Z_1} \tag{6.61}$$

$i_p < i_{t_1} < I$ where

$$L_s = \frac{a}{Z_1} e_{t_1} + b$$

and

$$R_s = \frac{e_{t_1}}{Z_1} + f$$

Equation (6.61) becomes:

$$2e_{t_1} = C Z_1 \frac{de_{t_2}}{dt} - e_{t_2} + \frac{Z_1}{Z_2} e_{t_2} - \frac{de_{t_1}}{dt} \left(\frac{a}{Z_1} e_{t_1} + \frac{b}{Z_1} \right) - e_{t_1} \left(\frac{c}{Z_1^2} e_{t_1} + \frac{f}{Z_1} \right) \tag{6.62}$$

Applying the process of Laplace transform on Eq. (6.62) to obtain the solution for e_{t_2}, and using the special rule outlined in Eqs. (6.34) and (6.35), and with zero initial conditions, the solution for $e_{t_2}(t)$ is given by:

$$e_{t_2}(t) = \frac{k_3 Z_2}{Z_1 - Z_1} + e^{-\frac{z_1 - z_2}{c z_1 z_2}} \left[\frac{k_4}{C Z_1} - \frac{k_3 Z_2}{Z_1 - Z_2} \right] \tag{6.63}$$

where

$$k_3 = \frac{A}{Z_1} \left[2Z_1 + f + C \frac{A}{Z_1} \right] \tag{6.64}$$

and

$$k_4 = \frac{A}{Z_1} \left[\frac{aA}{Z_1} + b \right] \tag{6.65}$$

or

$$k_3 = \left[2A + \frac{A}{Z_1} \frac{i_p R_{s-max}}{i_p - I} + \frac{A^2}{Z_1^2} \frac{R_{s-max}}{I - i_p} \right] G(x) \tag{6.66}$$

and

$$k_4 = \left[\frac{A^2}{Z_1^2} \frac{L_{s-max}}{I - i_p} + \frac{A}{Z_1} \frac{i_p L_{s-max}}{I - i_p} \right] G(x) \tag{6.67}$$

Inspection for k_3 and k_4 in Eqs. (6.66) and (6.67) shows clearly the high possibility of weaker e_{t_2} due to the saturable-resistor function in a region of rising protection.

Turning now to the region at which the saturable-resistor ohmic values start to decline toward minimum level, i.e.,

$$I < i(t) < I_{max}$$

At this range, L_s and R_s are rewritten here for reader convenience:

$$L_s = \frac{a'}{Z_1} e_{t_1} + b'$$

$$R_s = \frac{c'}{Z_1} e_{t_1} + f'$$

In a similar process used to solve for $e_{t_2}(t)$ for $i_p < i(t) < I$, the solution for $e_{t_2}(t)$ for the end phase of saturable-resistor behavior is given by:

$$e_{t_2}(t) = k_3' \frac{Z_2}{Z_1 - Z_1} + \left[\frac{k_4'}{c' Z_1} - \frac{k_3' Z_2}{Z_1 - Z_2} \right] e^{-\frac{z_1 - z_2}{c z_1 z_2} t} \tag{6.68}$$

where

$$k_3' = \frac{A}{Z_1}\left(2Z_1 + f' + C\,\frac{A}{Z_1}\right)G(x)$$ (6.69)

$$k_4' = \frac{A}{Z_1}\left(\frac{a'A}{Z_1} + b'\right)G(x)$$ (6.70)

Inserting a', b', c', and f' from Eqs. (6.10) and (6.11) into Eqs. (6.69) and (6.70) results in the following:

$$k_3' = \frac{A}{Z_1}\left[2Z_1 - \frac{I_{max}R_{s-max}}{I_{max} - I} - \frac{A}{Z_1}\,\frac{R_{s-max}}{I - I_{max}}\right]G(x)$$ (6.71)

$$k_4' = \frac{A}{Z_1}\left[\frac{A}{Z_1} - \frac{I_{max}R_{s-max}}{I_{max} - I} + A\,\frac{I_{max}L_{s-max}}{I_{max} - I}\right]G(x)$$ (6.72)

Quantitative comparison for k_3' and k_4' based on relative values of elements shown in Eqs. (6.71) and (6.72) indicates a convincing effect for the saturable-resistor behavior in weakening the level of the transmitted surge $e_{t_2}(t)$, but in a region of declining protection.

6.5 Equivalent Resistance of Saturable-Resistor Across Ideal Capacitor

In this section, the role of the equivalent resistance of the saturable-resistor is considered in a scheme of parallel connection with an ideal capacitor. R_{s-eq} is being considered as the equivalent resistance part of total ohmic value that has another parallel component, which is the inductive part.

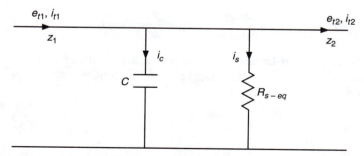

Figure 6.6 Shunt combination of ideal capacitor and saturistor equivalent resistance.

The circuit equations of continuity are:

$$i_{t_1} + i_{r_1} = i_c + i_s + i_{t_2} \tag{6.73}$$

or

$$\frac{e_{t_1}}{Z_1} - \frac{e}{Z_1} = c\frac{de_{t_2}}{dt} + i_s + \frac{e_{t_2}}{Z_2} \tag{6.74}$$

Therefore,

$$e_{t_1} - e_{r_1} = cZ_1\frac{de_{t_2}}{dt} + Z_1 i_s + \frac{Z_1}{Z_2} e_{t_2} \tag{6.75}$$

$$e_{t_1} + e_{r_1} = e_{t_2} \tag{6.76}$$

Adding Eqs. (6.75) and (6.76) results in:

$$2e_{t_1} = e_{t_2} + cZ_1\frac{de_{t_2}}{dt} + Z_1 i_s + \frac{Z_1}{Z_2} e_{t_2} \tag{6.77}$$

To express i_s in terms of e_{t_2}, we can write:

$$e_{t_1} = Ri_s = i_s(ai_s + b) \tag{6.78}$$

From Eq. (6.78), i_s is given by:

$$i_s = -\frac{b}{2a} \pm \frac{1}{2a}\sqrt{b^2 + 4ae_{t_1}} \tag{6.79}$$

To simplify i_s, but not to lose the principle behavior of the saturable-resistor, we are implementing the fact that b' is, practically, very small and therefore $b^2 \ll 4ae_{t_2}$.

Consequently, $i_s \approx \sqrt{4ae_{t_1}}/2a$.

Equation (6.77) becomes:

$$2e_{t_1} = e_{t_2} + CZ_1\frac{de_{t_2}}{dt} + \frac{Z_1}{Z_2} e_{t_2} \pm \frac{Z_1}{2a}\sqrt{4ae_{t_1}} \tag{6.80}$$

Applying the process of Laplace transform on Eq. (6.80) at zero initial conditions, and with $e_{t_1} = AU_{-1}(t)G(x)$ as the incident voltage surge e_{t_1}, the solution for $e_{t_2}(t)$ is given by:

$$e_{t_2} = \frac{Z_2 G(x)}{Z_1 + Z_2}\left[2A \mp sqrt\,\frac{A}{a}\right]\left[1 - e^{-\frac{z_1 + z_2}{cz_1 z_2}t}\right] \tag{6.81}$$

Equation (6.81) is valid for $i_p < i_s < I$.

The solution for $e_{t_2}(t)$ for $I < i_s < I_{max}$, and is given by:

$$e_{t_2} = \frac{Z_2 G(x)}{Z_1 + Z_2} \left[2A \mp sqrt \ \frac{A}{a'} \right] \left[1 - e^{-\frac{z_1 + z_2}{c z_1 z_2} t} \right] \qquad (6.82)$$

The level of $e_{t_2}(t)$ given by Eq. (6.82) is in a region where the saturable-resistor provides its major rising protective effect.

6.6 Equivalent Inductance of Saturable-Resistor Across Ideal Inductor

In this protective scheme, the equivalent shunt inductance of the saturable-resistor is connected in parallel with an ideal inductor.

Equations of circuit theory continuity are:

$$i_{t_1} + i_{r_1} = i_s + i_2 + i_{t_2} \qquad (6.83)$$

or

$$\frac{i_{t_1}}{Z_1} - \frac{e_{r_1}}{Z_1} = \frac{1}{L_s} \int e_{t_1} dt + \frac{1}{L} \int e_{t_2} dt + \frac{e_{t_2}}{Z_2} \qquad (6.84)$$

$$e_{t_1} - e_{r_1} = \frac{Z_1}{L_s} \int e_{t_1} \, dt + \frac{Z_1}{L} \int e_{t_2} \, dt + \frac{Z_1}{Z_2} e_{t_2} \qquad (6.85)$$

and $e_{t_1} + e_{r_1} = e_{t_2}$.
  $L_s = a i_s + b$ for $i_p < i_s(t) < I$ and

$$L_s \frac{di_s}{dt} = e_{t_1} \qquad (6.86)$$

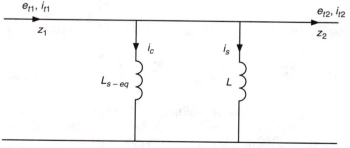

Figure 6.7 Shunt combination of ideal inductor and saturistor equivalent inductor.

Therefore,

$$(ai_s + b) \frac{di_s}{dt} = e_{t_1} = AU_{-1}(t)G(x) \tag{6.87}$$

or

$$\int (ai_s + b)di_s = G(x) \int AU_{-1}(t) \tag{6.88}$$

$$\frac{1}{2} ai_s^2 + bi_s = (At + \text{constant})G(x) \tag{6.89}$$

At $t = 0$, $i_s = 0$. Therefore the constant $= 0$. Therefore,

$$i_s = -\frac{b}{a} \pm \sqrt{\frac{b^2}{a^2} + \frac{2At}{a}} \tag{6.90}$$

For the same reasoning mentioned earlier, that $b^2/a^2 \ll 2At/a$ and that b/a is very small, relatively:

$$i_s \approx \sqrt{\frac{2At}{a}} \tag{6.91}$$

Returning to Eq. (6.85) for the time being, we will deal with the term:

$$\frac{Z_1}{ai_s + b} \int AU_{-1}(t)dt$$

$$\frac{Z_1 G(x)}{ai_s + B} \int AU_{-1}(t)dt = \frac{Z_1 t}{a\sqrt{\frac{2A}{a}} + b} \tag{6.92}$$

$$= \left[\frac{Z_1}{\gamma} t^{\frac{1}{2}} - \frac{bZ_1}{\gamma^2} + \cdots \right] \tag{6.93}$$

where $\gamma = \sqrt{2aA}$.

Now, inserting Eq. (6.93) into Eq. (6.85) and adding the result to Eq. (6.86) gives:

$$2e_{t_1} = \frac{Z_1}{L} \int e_{t_2}dt + \frac{Z_1}{Z_2} e_{t_2} + \frac{Z_1}{\gamma} \sqrt{t} - \frac{bZ_1}{\gamma^2} \tag{6.94}$$

Applying the process of Laplace transform on Eq. (6.94) and treating $e_{t_1} = AU_{-1}(t)G(x)$ such that the induced voltage surge on a transmission line is due to a sustained lightning surge, $e_{t_2}(t)$ is given by:

$$e_{t_2}(t) = \frac{m_1 Z_2}{Z_1} e^{-\frac{Z_2}{L}t} - \frac{m_2 Z_2}{Z_1} \left[\frac{2}{\sqrt{Z_2 \pi/L}} e^{-\frac{Z_2 t}{L}} \int_0^{\frac{Z_2 \sqrt{t}}{L}} e^{\lambda^2}d\lambda \right] \tag{6.95}$$

where

$$m_1 = \left(2A + \frac{bZ_1}{\gamma^2}\right) G(x)$$

$$m_2 = \frac{Z_1}{2\gamma} \sqrt{\pi} G(x) \tag{6.96}$$

Equation (6.95) is valid when $i_p < i_s(t) < I$ (i.e., in the region where the saturable-resistor ohmic values are below their peaks), providing a region of rising protection.

For the case where $I < i_s(t) < I_{max}$, Eq. (6.95) represents the sought solution, but replacing the constant b by f and a by c.

6.7 Surge Capture and Utilization

In this chapter, new devices or schemes involving connection of the saturable-resistor with either an ideal inductor or ideal capacitor have been analyzed to secure solutions for the weakened and stretched transmitted voltage surge into a second region of different surge impedance. All configurations indicate directly or indirectly a clear, effective role for the system or device in weakening the ultimate limit of the transmitted surge identified by $e_{t_2}(t)$. In Chap. 5, comprehensive presentation was given for the feasibility of using the captured surges $e_{t_2}(t)$ in exciting energy-producing converters through a controlled design process. The energy converters considered suitable are the electrochemical storage battery, the redox rechargeable flow cell, the process of electrolysis and electroseparation, the magnetohydrodynamic generators, and the electromechanical induction generator.

Similarly, the transmitted surges whose solutions have been presented in this chapter by devices containing the saturable-resistor in conjunction with ideal (L) or (c) could also be used to excite any of those energy converters through a controlled process. We have to emphasize the fact that $e_{t_2}(t)$ analyzed in this chapter was in reference to an incident and sustained induced surge on a transmission line only. A new $e_{t_2}(t)$ could be solved for if the incident surge is that induced at the entry of the transformer or transmission line tower. Therefore, this author would like to indicate clearly the added incentive of using the captured transmitted surges into these modes of energy conversion.

6.8 Problems

6.1 In reference to Fig. 6.2, obtain an expression for the induced transmitted voltage surge if the incident surge $e_{t_1}(t)$ is that developed at the entry of the transformer given by Eq. (4.30).

6.2 Repeat Prob. 6.1 with respect to Fig. 6.3 involving series connection of ideal capacitor with saturable-resistor.

6.3 Repeat Prob. 6.1 with respect to Fig. 6.4 involving a series saturable-resistor in line 1 and a shunt ideal across inductor.

6.4 Repeat Prob. 6.1 with respect to Fig. 6.5 involving a series saturable-resistor in line 1 and a shunt ideal across capacitor.

6.5 Repeat Prob. 6.1 with respect to Fig. 6.6 involving parallel combination of ideal capacitor and equivalent resistance of a saturable-resistor.

6.6 Repeat Prob. 6.1 with respect to Fig. 6.7 involving parallel combination of ideal inductor, ideal capacitor, and equivalent resistance of a saturable-resistor.

6.7 Repeat Prob. 6.1 with respect to a system involving parallel combination of ideal inductor, ideal capacitor, and equivalent resistance of a saturable-resistor.

6.8 Repeat Prob. 6.1 with respect to a system involving parallel combination of ideal inductor, ideal capacitor, equivalent resistance of a saturable-resistor, and equivalent inductance of saturable-resistor.

6.9 From the solution of $e_{t_2}(t)$ secured in Prob. 6.1, obtain the solution for the energy captured by the ideal inductor. Also, identify the condition for maximum energy stored with respect to parameters of the saturable-resistor.

6.10 From the solution of $e_{t_2}(t)$ secured in the text for $e_{t_1}(t)$ induced on a transmission line, obtain the expression for the energy stored in the ideal capacitor. Then identify the condition on the saturable-resistor's parameters for maximum energy stored.

6.11 In a system involving a series saturable-resistor and a shunt ideal capacitor, using the calculated $e_{t_2}(t)$ obtained in this chapter, secure the solution for the energy stored in the capacitor. Then identify conditions for maximum energy stored with respect to parameters of the saturable-resistor.

6.12 In reference to the system shown in Fig. 6.6, using the solution obtained in this chapter for $e_{t_2}(t)$ due to induced surge on a transmission line, obtain the expression for the energy stored in the capacitor. Then identify conditions on the parameters of the saturable-resistor for maximum energy stored in the capacitor.

6.9 References

1. Alger, P. L., *Induction Machines,* 2d edition, Gordon and Breach Science Publishers, New York, London, Paris, 1970.

2. Alger, P. L., Angst, G., and Schweder, "Saturistor and Low Starting Current Induction Motors," *IEEE Trans. Power Apparatus & Systems,* vol. 82, June 1963, pp. 291–297.
3. Denno, K., *Power System Design and Applications for Alternative Energy Sources,* Prentice-Hall, Inc., Englewood Cliffs, N. J., 1989.
4. Denno, K., *Engineering Economics of Alternative Energy Sources,* CRC Press, Boca Raton, Florida, 1992.
5. Denno, K., "Eddy-Current Theory in Hard/, Thick Ferromagnetic Materials," IEEE Power Engineering Society Conference Paper C75-005-4, presented at the winter meeting in New York, 1975.
6. Denno, K., "Current Limiting in High Voltage Transmission System," *Proceedings of the Canadian Communications and EHV Conference,* 1972, pp. 155–156.
7. Gunn, C. E., "Improved Starting Performance of Wound-Rotor Motors Using Saturistors," *IEEE Trans. Power Apparatus & Systems,* vol. 82, June 1963, pp. 298–302.
8. Nilson, J. W., *Electric Circuit,* 3d edition, Addison Wesley Publishing Co., Reading, Massachusetts, 1990.
9. Rüdenberg, Reinhold, *Electrical Shock Waves in Power Systems,* Harvard University Press, Cambridge, Massachusetts, 1968.

Nonlinear Arresters with Saturable-Resistors, Inductances, and Capacitors

In Chap. 6, protective functions of saturable-resistors, inductors, and capacitors have been considered in various network design arrangements, such as series combination of inductors with saturable-resistors, series capacitors and saturable-resistors, shunt combination of saturable-resistor with series capacitor, and shunt capacitor with series saturable-resistor. The role of the saturable-resistor has been focused on its behavior as a series current arrester, since its ohmic values for the resistance and reactance components rise immediately to a much higher level as soon as the surge current exceeds the limit of the saturable-resistor pickup current. However, when the surge current surpasses the case at which the saturable-resistor attains its maximum ohmic value (i.e., magnetic aturation), the reverse behavior commences, where the ohmic values decline toward very small values for higher flows of surge currents. Therefore, we can see that the saturable-resistor, in effect, has a limited role as a surge current arrester, and other protective mechanisms ought to be devised to supplement the saturable-resistor for higher destructive induced current on transmission lines, transformers, towers, and any part of the power system AC grid network.

In this chapter, voltage surge arresters and current surge arresters in conjunction with conventional resistive dampers, together with ideal inductors and ideal capacitors, will be considered for design analysis and feasibility in performing the added function of energy capture and, of course, for eventual utilization.

The voltage surge arrester called for in this chapter is a nonlinear device that will allow very limited current to flow when the surge voltage level is of the order of the steady-state normal operating voltage, after which the current flow will be according to a predesigned mode of change which, in most cases, is linear, as shown in Fig. 7.1.

The current surge arrester is considered to be a special mode of nonlinear device designed on the basis of absorbing the main thrust of the current surge through a fully ionized gas column, and then, upon reduction in the current surge, deionization with ion recombination will render the gas column of much higher resistance, blocking further passage of any current up to the point where it is of the order of the normal operational level at which the spark switch opens. Figure 7.2 represents the current-voltage characteristic of the nonlinear current arrester.

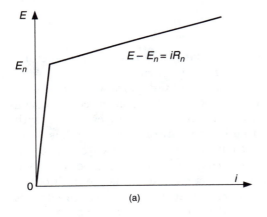

Figure 7.1 (*a*) Voltage surge arrester.

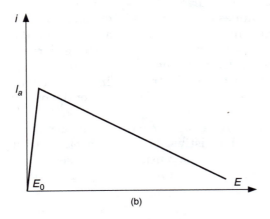

Figure 7.1 (*b*) Current surge arrester.

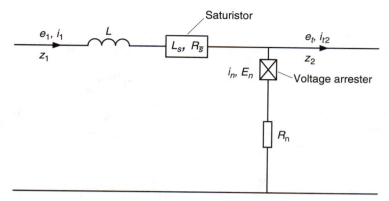

Figure 7.2 Protective combination system.

Therefore, the main features of the current arrester are characterized as high current flow through very low resistance created by the
highly ionized gas column and, at a relatively normal operating current, a high resistance settled by the deionized gas column will prevent
any current flow by the opening of the spark-gap switch.

The strategy in this chapter is to outline the key equations and physical realities for each potential design of protective system, listing the
expected outcome of each element, and then leaving completion of the
mathematical solution in detail as a problem for the student.

7.1 Ideal Inductor, Saturistor, and Voltage Arrester

This is shown in Fig. 7.2.

Continuity equations based on circuit theory concepts for the mechanism shown in Fig. 7.2 are:

$$e_{t_1} + e_{r_1} = L \frac{di_{t_1}}{dt} + L_s \frac{di_{t_1}}{dt} + R_s i_{t_1} + e_{t_2} \tag{7.1}$$

$$i_1 = i_2 + i_n \tag{7.2}$$

$$i_n = \frac{e_{t_2} - E_n}{R_n} \tag{7.3}$$

Therefore, from Eqs. (7.2) and (7.3):

$$\frac{e_{t_1}}{Z_1} - \frac{e_{r_1}}{Z_1} = \frac{e_{t_2}}{Z_2} + \frac{e_{t_2} - E_n}{R_n} \tag{7.4}$$

or

$$e_{t_1} - e_{r_1} = \frac{Z_1}{Z_2} e_{t_2} + \frac{Z_1}{R_n}(e_{t_2} - E_n) \tag{7.5}$$

Equation (7.1) could be written in another form as follows:

$$e_{t_1} + e_{r_1} = \frac{L}{Z_1} \frac{de_{t_1}}{dt} + \frac{L_s}{Z_1} \frac{de_{t_1}}{dt} + \frac{R_s}{Z_1} e_{t_1} + e_{t_2} \tag{7.6}$$

Adding Eqs. (7.5) and (7.6) gives:

$$2e_{t_1} = \frac{L}{Z_1} \frac{de_{t_1}}{dt} + \frac{L_s}{Z_1} \frac{de_{t_1}}{dt} + \frac{R_s}{Z_1} e_{t_1} + e_{t_2}\left(1 + \frac{Z_1}{R_n}\right) - \frac{Z_1 E_n}{R_n} \tag{7.7}$$

where

E_n is the order of the normal operating voltage of the operating power system.

R_n is the linear resistance of the voltage surge arrester described by the relationship $e_{t_1} - E_n = iR_n$.

Also, from Eqs. (6.1) and (6.2), L_s and R_s before saturation could be written as:

$$L_{s-before} = ai_{t_1}(t) + b, \text{ for } i_p < i_{t_1} < I_{max}$$
$$R_{s-before} = ci_{t_1}(t) + f, \text{ for } i_p < i_{t_1} < I_{max} \tag{7.8}$$

while Eqs. (6.10) and (6.11) describe L_s and R_s after saturation, as rewritten here:

$$L_{s-after} = a'\hat{\imath}_{t_1}(t) + b', \text{ for } I_{max} < i_{t_1} < I_{sat}$$
$$R_{s-after} = c'\hat{\imath}_{t_1}(t) + f', \text{ for } I_{max} < i_{t_1} < I_{sat} \tag{7.9}$$

The solution of the transmitted voltage and current surges from Eq. (7.8), in correlation with Eqs. (7.9) and (7.10), depends upon the nature of the incident or induct voltage surge e_{t_1}.

e_{t_1} could be the induced voltage on a transmission line due to a sustained lightning surge, or at the entry of a transformer, a switching singularity surge, oscillatory surge, or any other arbitrary waveform surge.

Potential benefits expected from the network shown in Fig. 7.2 are:

1. The ideal inductor will flatten the incoming surge, storing an amount of magnetic energy equal to $\frac{1}{2}LI^2_{final}$ joules.

2. The immediate rise in the saturable-resistor impedance will act as a current arrester up to the limit when $i_{t_1} = I_{max}$. (I_{max} is the current flow in the saturable at which its impedance is maximum.)

3. The saturable-resistor resistance component will act as an effective damper.

4. The voltage arrester will weaken e_{t_2} by the limit of E_n. (E_n is usually the normal operating voltage of the system.)

5. The transmitted current surge i_{t_2} will be weakened according to this relationship:

$$i_{t_2} = e_{t_2} - E_n/Z_2 \tag{7.10}$$

7.2   Ideal Capacitor, Saturistor, and Voltage Surge Arrester

This is shown in Fig. 7.3, where the ideal capacitor in series with a saturable-resistor and a voltage arrester having a spark limit of normal operating voltage is connected across the line.

Continuity equations on circuit theory for the configuration of Fig. 7.3 are:

$$e_{t_1} + e_{r_1} = \frac{1}{c} \int i_{t_1} dt + L_s \frac{di_{t_1}}{dt} + R_s i_{t_1} + e_{t_2} \tag{7.11}$$

or

$$e_{t_1} + e_{r_1} = \frac{1}{cZ_1} \int i_{t_1} dt + \frac{L_s}{Z_1} \frac{di_{t_1}}{dt} + \frac{R_s}{Z_1} e_{t_1} + e_{t_2} \tag{7.12}$$

$$i_1 = i_n + i_2 \tag{7.13}$$

$$\frac{e_{t_1}}{Z_1} - \frac{e_{r_1}}{Z_1} = \frac{e_{t_2} - E_n}{R_n} + \frac{e_{t_2}}{Z_2}$$

$$e_{t_1} - e_{r_1} = \frac{Z_1}{R_n}(e_{t_2} - E_n) + \frac{Z_1}{Z_2} e_{t_2} \tag{7.14}$$

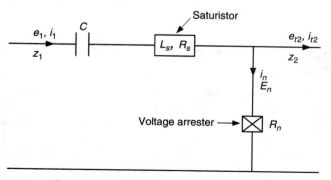

Figure 7.3 Protective combination system.

Adding Eqs. (7.12) and (7.14) gives:

$$e_{t_1} - \frac{1}{2cZ_1} \int e_{t_1} dt - \frac{L_s}{Z_1} \frac{de_{t_1}}{dt} - \frac{R_s}{Z_1} e_{t_1}$$

$$= e_{t_2}\left(1 + \frac{Z_1}{Z_2} + \frac{Z_1}{R_n}\right) - \frac{Z_1 E_n}{R_n} \qquad (7.15)$$

e_{t_1} in the preceding equation is the voltage surge induced on a transmission line which could be due to a sustained lightning surge, actual lightning surge, voltage surge induced at the entry of transformer, tower, and all other kinds of switching surges.

Inserting e_{t_1} in Eq. (7.15), together with the mathematical descriptions for L_s and R_s before and after magnetic saturation, the solution for e_{t_2} will emerge again before and after magnetic saturation through the solution of this differential equation. Solutions for e_{t_2} are left to the student, since similar cases have been dealt with in previous chapters.

Expectations for the potential protective and energy storage role elements shown in Fig. 7.3 are:

1. The ideal capacitor will capture electrostatic energy in the amount of $Q^2/2c$. (Q is the electric charge stored.)

2. The saturable-resistor impedance will rise as the surge current i_{t_1} increases, in effect limiting a further rise in i_{t_1}. The resistive component of the saturable-resistor will act as a surge damper.

3. Current i_{t_2} will be reduced by virtue of the role of the voltage arrester according to the criterion:

$$i_{t_1} = \frac{e_{t_2} - E_n}{R_n}$$

and of course the voltage surge e_{t_2} will be, effectively, much weakened.

7.3 Current Arrester with Ideal Inductor and Saturable-Resistor

This protective and energy storage device is shown in Fig. 7.4. The current arrester function is modeled as shown in Fig. 7.2, where a high current inrush will flow through the ionized gas column characterized by low ionization potential at the declining portion of the surge current i_n, which is just equal to or less than the normal leakage current at which the spark gap switch will open.

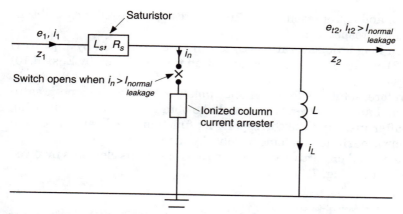

Figure 7.4 Protective combination system.

Continuity equations based on the principles of circuit theory are:

1. When $i_{t_1} \geq I_{normal}$, then

$$e_{t_1} + e_{r_1} = L_s \frac{di_{t_1}}{dt} + R_s i_{t_1} = \frac{L_s}{Z_1} \frac{de_{t_1}}{dt} + \frac{R_s}{Z_1} e_{t_1} \qquad (7.16)$$

2. When $i_{t_1} < I_{normal}$, then

$$e_{t_1} + e_{r_1} = L_s \frac{di_{t_1}}{dt} + R_s i_{t_1} + e_{t_2} \qquad (7.17)$$

$$i_{t_1} + i_{r_1} = i_{t_2} + i_2 \qquad (7.18)$$

where I is the normal current surge or normal leakage current at which the spark-gap switch opens and, consequently, i_{t_2} will flow toward the ideal inductor and the line.
Equations (7.17) and (7.18) become:

$$e_{t_1} + e_{r_1} = \frac{L_s}{Z_1} \frac{de_{t_1}}{dt} + \frac{R_s}{Z_1} e_{t_1} + e_{t_2} \qquad (7.19)$$

$$e_{t_1} - e_{r_1} = \frac{Z_1}{Z_2} e_{t_2} \qquad (7.20)$$

Adding these two equations gives:

$$2e_{t_1} = \frac{L_s}{Z_1} \frac{de_{t_1}}{dt} + \frac{R_s}{Z_1} e_{t_1} + e_{t_2} \left(1 + \frac{Z_1}{Z_2}\right) \qquad (7.21)$$

where L_s and R_s represent the saturable-resistors' resistance and self-inductance, respectively.

Before magnetic saturation, L_s and R_s are expressed by Eqs. (6.1) and (6.2), while after magnetic saturation they are given by Eqs. (6.10) and (6.11).

Therefore, solutions for e_{t_2} will be twofold: one before magnetic saturation and another after, which could be obtained from solution of Eq. (7.23) after inserting the corresponding functions for L_s and R_s. Those solutions are left as problems to solve by the student.

The expected protective and storing roles of various elements involved in the device of Fig. 7.4 are as follows:

1. When $i_{t_1} > I_{normal-leakage}$
   a. Inrush surge current will be shunted by the swiftly ionized gas column of the surge arrester.
   b. The rise in the saturable-resistor impedance will impose effective limitation on the rise of surge current, thereby allowing gradual passage through the ionizing surge arrester.
2. When $i_{t_1} \leq I_{normal-leakage}$. In this period, the gas column in the arrester will go through the process of deionization and ion recombination and, in effect, build a high resistive path, preventing passage of almost any part of the remainder of the surge current. Therefore, i_{t_1} will change route to the ideal inductor and to zone 2 of the power system having a surge impedance of Z_2.

   Expected protective and energy-storing functions of elements shown in Fig. 7.4 are as follows:
   a. Impedance of the saturable-resistor will decline drastically, allowing normal and smooth flow of i_{t_1} to the inductor and line 2.
   b. A modest amount of energy equal to $\frac{1}{2}LI_L^2$ joules will be stored in the ideal inductor.
   c. Substantially weakened surge current and voltage will flow in the section having surge impedance Z_2.

7.4   Saturable-Resistor, Ideal Capacitor, and Voltage Arrester

This is shown in Fig. 7.5.

The volt-ampere characteristic for the voltage surge is shown in Fig. 7.1a. Continuity equations for the network of Fig. 7.5 are:

$$e_{t_1} + e_{r_1} = L_s \frac{di_{t_1}}{dt} + R_s i_{t_1} + e_{t_2} \qquad (7.22)$$

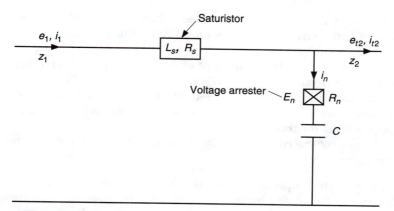

Figure 7.5 Protective combination system.

or

$$e_{t_1} + e_{r_1} = \frac{L_s}{Z_1} \frac{di_{t_1}}{dt} + \frac{R_s}{Z_1} e_{t_1} + e_{t_2} \qquad (7.23)$$

$$i_{t_1} + i_{r_1} = i_n + i_{t_2} \qquad (7.24)$$

or

$$\frac{e_{t_1}}{Z_1} - \frac{e_{r_1}}{Z_1} = \frac{e_{t_2}}{Z_2} + \frac{1}{c} \int i_n dt \qquad (7.25)$$

where

$$i_n = \frac{e_{t_2} - E_n}{R_n} = \frac{e_{t_2}}{R_n} - \frac{E_n}{R_n} \qquad (7.26)$$

Therefore, Eq. (7.25) becomes:

$$e_{t_1} - e_{r_1} = \frac{Z_1}{Z_2} e_{t_2} + \frac{Z_1}{c} \int \left(\frac{e_{t_2}}{R_n} - \frac{E_n}{R_n} \right) dt \qquad (7.27)$$

Now, adding Eqs. (7.23) and (7.27) gives:

$$2e_{t_1} = \frac{L_s}{Z_1} \frac{de_{t_1}}{dt} + \frac{R_s}{Z_1} e_{t_1} + e_{t_2} \left(1 + \frac{Z_1}{Z_2} \right) + \frac{Z_1}{cR_n} \int e_{t_2} dt - \frac{Z_1}{cR_n} E_n dt \qquad (7.28)$$

For the region below magnetic saturation, L_s and R_s have been given by Eqs. (6.1) and (6.2), while for those beyond magnetic saturation, L_s and R_s have been given by Eqs. (6.10) and (6.11).

Also, we have to reidentify that E_n is the voltage across the arrester at which very little current will pass and, after that, $(e_{t_2} - E_n)$ is proportional linearly with respect to i_n.

Solutions for e_{t_2} and i_{t_2} after short-circuit of the voltage-gap arrester could be obtained from solving the differential equation of Eq. (7.28), after substituting the functional forms for L_s and R_s, before and after magnetic saturation. The challenge of obtaining solutions for e_{t_2} and i_{t_2} are left as problems for the reader to solve.

Potential roles by the components devised in Fig. 7.5 for protection and energy storage are as follows:

1. Saturable-resistor impedance will rise sharply with the flow of inrush current generated by the induced voltage surge e_{t_1}, thereby limiting the current surge and weakening e_{t_2}. (This is before magnetic saturation.)

2. With e_{t_2} already weakened by the voltage drop through the saturable-resistor, the spark of the voltage arrester will remain open as long as $e_{t_2} < E_n$, after which it will flow through the ideal capacitor.

3. After magnetic saturation in the saturable-resistor, its impedance will diminish drastically, and if the surge voltage is in excess of E_n, i_n will flow through the ideal capacitor.

4. Whenever $e_{t_2} > E_n$, the ideal capacitor will store electrostatic energy of $Q^2/2C$ joules, where Q is the charge residing in the capacitor plates.

5. The protective combination of Fig. 7.5 is represented by the behavior of the saturable-resistor as current limiter and the voltage surge arrester as voltage limiter.

7.5 Current Arrester across Ideal Capacitor-Saturable-Resistor in Series with Zone 2

The protective network of Fig. 7.6 basically gives a major role for the current surge arrester to short-circuit the induced excessive current surge i_{t_1} whenever it is larger than the normal leakage current of the system I_{n-1}. Also, the network features the series location of the saturable-resistor in zone 2, to act in weakening or limiting the surge current i_{t_2} into that zone, a situation that may arise if i_{t_2} is still at an undesirable limit, or just in case the current surge arrester fails to perform its short-circuiting role (such as the stalling of the ionization process in its gas column).

Equations of continuity based on circuit theory for the network of Fig. 7.6 are as follows:

1. $i_n \geq I_{normal}$

$$e_{t_1} + e_{r_1} = 0 \qquad (7.29)$$

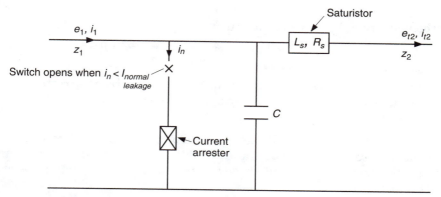

Figure 7.6 Protective combination system.

However, Eq. (7.29) will remain valid for a very short time, on the order of 100 ~300 μsec.

2. $i_n < I_{normal}$. In this situation, the current surge arrester switch will open, since the gas column will return to a deionizing state.

Circuit theory equations are:

$$e_{t_1} + e_{r_1} = \frac{1}{c} \int e_{t_1} dt + L_s \frac{di_{t_2}}{dt} + R_s i_{t_2} \qquad (7.30)$$

or

$$e_{t_1} + e_{r_1} = \frac{1}{c} \int e_{t_1} dt + \frac{L_s}{Z_2} \frac{de_{t_2}}{dt} + \frac{R_s}{Z_2} e_{t_2} \qquad (7.31)$$

$$e_{t_1} - e_{r_1} = \frac{de_{t_2}}{dt} + \frac{Z_1}{Z_2} e_{t_2} \qquad (7.32)$$

Adding Eqs. (7.31) and (7.32) gives:

$$2e_{t_1} = \frac{1}{c} \int e_{t_1} dt + cZ_1 \frac{de_{t_1}}{dt} + \frac{Z_1}{Z_2} e_{t_2} + \frac{L_s}{Z_2} \frac{de_{t_2}}{dt} + \frac{R_s}{Z_2} e_{t_2} \qquad (7.33)$$

Again, solutions for e_{t_2} and i_{t_2} could be obtained from Eq. (7.32) under two regimes: (1) below the limit of magnetic saturation where $i_p < i_{t_2} < I$ and (2) beyond magnetic saturation where $I < i_{t_2} < I_{max}$.

For the region below magnetic saturation, L_s and R_s are given by Eqs. (6.1) and (6.2), while for the region beyond magnetic saturation, L_s and R_s are given by Eqs. (6.10) and (6.11). Mathematical procedure to solve for e_t2 and i_t2 is left to the student.

Functional performance involving protection and energy storage for the network of Fig. 7.6 is centered on two stages:

a. $i_n \geq I_{normal-leakage}$.   The bulk of current surge will flow through the gas ionized column as a short-circuiting path.

b. $i_n \leq I_{normal-leakage}$.   The spark switch of the current arrester will open, at which point e_{t_1} will be subjected to the capacitor and i_{t_2} will flow in region 2.

The ideal condenser will store an amount of electrostatic energy equal to $\frac{1}{2}ce_{t_1}^2$ joules, while the saturable-resistor will act to further limit the remaining current surge i_{t_2} by virtue of the automatic rise of its impedance.

7.6 Ideal Capacitor, Saturable-Resistor, and Current Arrester

This is shown in Fig. 7.7, where an ideal capacitor is in series with a saturable-resistor, and a current arrester in series with ideal inductor. The saturable-resistor will impede the flow of the surge current i_{t_1} by the automatic rise of its impedance; besides, its resistance performs as an effective damper. Ideal inductor is in series with a current surge arrester.

Continuity equations based on circuit theory for the network shown in Fig. 7.7 are:

1. $i_n \geq I_{normal-leakage}$.   The current surge arrester will short-circuit the bulk of i_{t_1} through its ionized gas column. Therefore,

$$e_{t_1} + e_{r_1} = \frac{1}{c} \int i_{t_1} dt + L_s \frac{di_{t_1}}{dt} + R_s i_{t_1}$$

$$= 0 \tag{7.34}$$

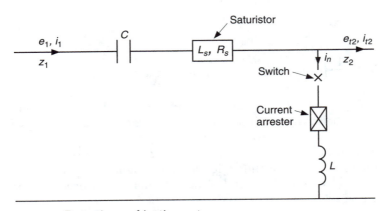

Figure 7.7 Protective combination system.

$$i_{t_1} + i_{r_1} = i_n = i_L \tag{7.35}$$

or

$$e_{t_1} - e_{r_1} = \frac{Z_1}{L} \int e_{t_2} dt \tag{7.36}$$

Also, Eq. (7.34) is rewritten as:

$$e_{t_1} + e_{r_1} = \frac{1}{cZ_1} \int e_{t_1} dt + \frac{L_s}{Z_1} \frac{de_{t_1}}{dt} + \frac{R_s}{Z_1} e_{t_1} \tag{7.37a}$$

$$e_{t_1} + e_{r_1} = 0 \tag{7.37b}$$

Adding Eqs. (7.36) and (7.37) gives:

$$2e_{t_1} = \frac{Z_1}{L} \int e_{t_2} \tag{7.38}$$

and

$$\frac{1}{cZ_1} \int e_{t_1} dt + \frac{L_s}{Z_1} \frac{de_{t_1}}{dt} + \frac{R_s}{Z_1} e_{t_1} = 0 \tag{7.39}$$

2. $i_n \leq I_{normal\text{-}leakage}$. At this stage, the surge current i_{t_1} is of the order of the normal leakage system current $I_{normal\text{-}leakage}$, at which point the spark switch of the current arrester opens and, consequently, i_{t_1} now proceeds into zone 2, whose surge impedance is Z_2.

Circuit theory equations for the network shown in Fig. 7.7 are:

$$e_{t_1} + e_{r_1} = \frac{1}{c} \int i_{t_1} dt + L_s \frac{di_{t_1}}{dt} + R_s i_{t_1} + e_{t_2} \tag{7.40}$$

or

$$e_{t_1} + e_{r_1} = \frac{1}{cZ_1} \int e_{t_1} dt + \frac{L_s}{Z_1} \frac{de_{t_1}}{dt} + \frac{R_s}{Z_1} e_{t_1} + e_{t_2} \tag{7.41}$$

Also,

$$i_{t_1} + i_{r_1} = i_{t_2} \tag{7.42}$$

or

$$e_{t_1} - e_{r_1} = \frac{Z_1}{Z_2} e_{t_2} \tag{7.43}$$

Now, adding Eqs. (7.41) and (7.43) gives:

$$2e_{t_1} = \frac{1}{cZ_1} \int e_{t_1} dt + \frac{L_s}{Z_1} \frac{de_{t_1}}{dt} + \frac{R_s}{Z_1} e_{t_1} + e_{t_2}\left(1 + \frac{Z_1}{Z_2}\right) \tag{7.44}$$

The solution for e_{t_2} when $i_n \leq I_{normal-leakage}$, as indicated in Eq. (7.37a), depends upon the functional form of e_{t_1}, which could be the induced voltage surge on a transmission line due to a sustained lightning surge, actual lightning surge, or a voltage induced at the entry of the transformer or tower.

Equation (7.37b) could be used for evaluating current limitation distribution through the ideal capacitor and the saturable-resistor ohmic values.

The solution for e_{t_2} and i_{t_2} when $i_n \leq I_{normal-leakage}$ could be obtained by solving the differential equation shown as Eq. (7.40), again depending upon the functional nature of the voltage surge e_{t_1}. Again, the solution for e_{t_2} is based on two regimes with respect to the state of magnetization of the saturable-resistor, namely, before and after magnetic saturation. Mathematical forms for L_s and R_s before and after saturation are given by Eqs. (6.1), (6.2), (6.10), and (6.11), respectively. Solutions are left as problems for the student to perform.

Turning now to identify the roles of protection and energy storage for the network of Fig. 7.7, we can attribute the role of the capacitor as surge current by virtue of the rise in its ohmic values, as well as the damping effect of its resistance.

However, when $i_n = i_L \geq I_{normal-leakage}$, the bulk of the surge current flows through the arrester and the ideal inductor, and an amount of energy equal to $\frac{1}{2}LI^2_{final}$ will be stored. Next, when $i_n \leq I_{normal-leakage}$, the spark switch of the current arrester opens (since the gas column of the arrester will turn to be deionized) and, consequently, the limited i_{t_1} proceeds to flow into zone 2, having a surge impedance of Z_2.

7.7   Current and Voltage Arresters with Saturable-Resistor and Ideal Inductor and Capacitor

Networks comprising these elements are shown in Fig. 7.8, where the saturable-resistor limits and, in effect, truncates the peak of i_{t_1}, while the ideal inductor captures the bulk amount of magnetic energy captured by the current arrester and, at the same time, the ideal capacitor will store electrostatic energy in the amount of $Q^2/2C$ joules after passing through the voltage arrester.

Consider first the incidence of current surge e_{t_1} at the zone of surge impedance Z_1, such that $i_{t_1} \geq I_{normal-leakage}$, at which point the current arrester will absorb i_L and, through the ideal inductor, store $\frac{1}{2}LI^2_{final}$ joules.

Circuit theory equations for this case are:

$$e_{t_1} + e_{r_1} = L_s \frac{di_{t_1}}{dt} + R_s i_{t_1} + L \frac{di_{t_1}}{dt} \tag{7.45}$$

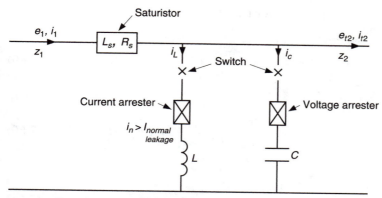

Figure 7.8 Protective combination system.

or

$$e_{t_1} + e_{r_1} = \frac{L_s}{Z_1}\frac{de_{t_1}}{dt} + \frac{R_s}{Z_1}e_{t_1} + \frac{L}{Z_1}\frac{de_{t_1}}{dt} \tag{7.46}$$

$$e_{t_1} - e_{r_1} = \frac{Z_1}{L}\int e_{t_2}dt \tag{7.47}$$

Adding Eqs. (7.46) and (7.47) gives:

$$2e_{t_1} = \frac{R_s}{Z_1}e_{t_1} + \frac{de_{t_1}}{dt}\left(\frac{L_s}{Z_1} + \frac{L}{Z_1}\right) + \frac{Z_1}{L}\int e_{t_2}dt \tag{7.48}$$

Solutions for e_{t_2} and i_2 could be secured below and beyond magnetic saturation by inserting the proper representation for L_s and R_s mentioned in the preceding sections of this chapter.

The present phase regarding the incidence of current surge will move now into the stage where $i_{t_1} \le I_{normal-leakage}$, at which point current flow will cease to flow through the ideal inductor. However, the remainder of current surge, which could be at a substantial voltage level, will flow into the voltage arrester if it is higher than the arrester nominal voltage threshold. If the voltage of remainent current surge is below E_n, nothing will flow into zone 2.

Second, let us consider the incidence of a voltage surge associated with electric current of the order of normal leakage level. In this case the current arrester will be idle while the voltage arrester swings into action; where $e_{t_2} \ge E_n$ (E_n is the threshold level for the voltage arrester), i_c will flow and an amount of electrostatic energy equal to $Q^2/2C$ joules will be stored in the capacitor and, at the same, the surge e_{t_2} will release i_{t_2} into zone 2.

Circuit theory equations for this case involving the incidence of voltage surge are:

$$e_{t_1} + e_{r_1} = L_s \frac{di_{t_1}}{dt} + R_s i_{t_1} + e_{t_2} \tag{7.49}$$

or

$$e_{t_1} + e_{r_1} = \frac{L_s}{Z_1} \frac{de_{t_1}}{dt} + \frac{R_s}{Z_1} e_{t_1} + e_{t_2} \tag{7.50}$$

and

$$i_{t_1} + i_{r_1} = i_{t_2} + \frac{e_{t_2} - E_n}{R_n} \tag{7.51}$$

or

$$e_{t_1} - e_{r_1} = \frac{Z_1}{Z_2} e_{t_2} + \frac{e_{t_2}}{R_n} - \frac{E_n}{R_n} \tag{7.52}$$

Adding Eqs. (7.50) and (7.52) gives:

$$2e_{t_1} = \frac{R_s}{Z_1} e_{t_1} + \frac{L_s}{Z_1} \frac{de_{t_1}}{dt} + e_{t_2} (1 + \frac{Z_1}{Z_2} - \frac{1}{R_n}) - \frac{E_n}{R_n} \tag{7.53}$$

The solution for e_{t_2} as well as i_{t_2} could be secured from Eq. (7.46) below and beyond magnetic saturation regarding the behavior of the saturable-resistor after inserting the proper representation for L_s and R_s. However, the most likely situation is that the current associated with the voltage surge may be of the order of $I_{normal-leakage}$, at which the role of the saturable-resistor will be minimal, and the main protective role played is that by the voltage arrester and its resistance.

Solutions for the transmitted current and voltage surges from Eqs. (7.43) and (7.46) depends upon the nature of the incident surge, and are left as problems for the reader.

Discussing the functional roles of protection and energy storage for the mechanism shown in Fig. 7.8, first in the phase where a current surge is incident on zone 1, the saturable-resistor automatic rise in impedance will limit the rise of the surge when $i_{t_1} > I_{normal-leakage}$, followed by the action of the surge arrester that will allow the passage of bulk surge current which finally will be stored in the ideal inductor. With the incidence of a voltage surge, its current content may be weakened by the saturable-resistor. The current arrester will not act if $i_{t_1} \leq I_{normal-leakage}$; however, the voltage arrester will step into action with a certain amount of electrostatic energy being stored in the capacitor.

7.8 Problems

7.1 Fig. 7.1*a* represents the volt-ampere characteristic for the voltage arrester, while that of Fig. 7.1*b* represents that for the current-surge arrester. Derive an appropriate mathematical function for each.

7.2 Solve differential Eq. (7.8) for e_{t_2} when e_{t_1} is the induced voltage on a transmission line due to a sustained lightning surge. Assume zero initial conditions.

7.3 Repeat Prob. 7.2 if e_{t_1} is the induced voltage surge on a transmission line due to an actual lightning surge characterized by a sharp front rise and a relatively slow decaying tail.

7.4 Repeat Prob. 7.2 if e_{t_1} is the voltage surge induced at the entry of the transformer substation due to a sustained lightning surge.

7.5 In reference to the differential equation of Eq. (7.15), solve for e_{t_2} and i_{t_2}, where e_{t_1} is the voltage surge induced on a transmission line due to a sustained lightning surge. Assume zero initial conditions.

7.6 Repeat Prob. 7.5 if i_{t_1} is the current surge induced on a transmission line due to an actual lightning stroke.

7.7 Repeat Prob. 7.5 if i_{t_1} is the current surge induced at the entry of a transformer substation due to a sustained lightning stroke. Assume zero initial conditions.

7.8 In reference to the differential equation of Eq. (7.23), solve for e_{t_2} and i_{t_2}, where e_{t_1} is the voltage surge induced on a transmission line due to actual lightning stroke. Assume zero initial conditions.

7.9 Repeat Prob. 7.8 if e_{t_1} is the voltage surge induced at the entry of a transformer substation is generated by a sustained lightning stroke.

7.10 Repeat Prob. 7.8 if e_{t_1}, the induced voltage surge, is a periodic sinnsoidal train. Assume zero initial conditions.

7.11 In reference to the differential equation of Eq. (7.28), solve for e_{t_2} and i_{t_2}, where i_{t_1} is the induced current surge due to a sustained lightning stroke on a transmission line. Assume zero initial conditions.

7.12 In reference to differential equation of Eq. (7.32), solve for e_{t_2} and i_{t_2}, where e_{t_1} is the voltage surge induced on a transmission line due to an actual lightning stroke. Assume zero initial conditions.

7.13 Repeat Prob. 7.12 if e_{t_1} is the induced voltage surge at the entry of a transformer due to a sustained lightning surge.

7.14 In reference to differential equation of Eq. (7.40), solve for e_{t_2} and i_{t_2}, where i_{t_1} is the current surge on a transmission line due to a sustained lightning surge. Assume zero initial conditions.

7.15 In reference to differential equation of Eq. (7.46), solve for e_{t_2} and i_{t_2}, where i_{t_1} is the induced current surge in the form of a sinnsoidal periodic train in the time domain only.

7.16 Repeat Prob. 7.15 if e_{t_1} is the voltage surge induced at the entry of a transformer due to a sustained lightning surge. Assume zero initial conditions.

7.9 References

1. Cobine, J. D., *Gaseous Conductors (Theory Engineering Applications)*, Dover Publications, Inc., New York, 1941, 1958.
2. Denno, K., *High Voltage Engineering in Power Systems*, CRC Press, Boca Raton, Florida, 1992.
3. Rüdenberg, Reinhold, *Electrical Shock Waves in Power Systems*, Harvard University Press, Cambridge, Massachusetts, 1968.

8

Symmetrical Components

8.1 Introduction

In this chapter, a generalized review of the principles and applications of symmetrical components will be presented.

The review process will focus mainly on outlining the basic principles of symmetrical components, the transformation matrix from phase sequence to symmetrical sequence, the mode of interconnection of sequence network at any fault, and the practice of using the bus-impedance system.

This subject matter is to be used in the process of coordination of polarity power relays as well as in the analysis of every unbalanced power system which involves determination of sequence voltage, current, and power in a power system subjected to asymmetrical faults.

Three-phase short circuit, which is the most severe fault, keeps the system totally balanced, while other kind of faults such as line-ground fault, line-line fault or double-line-ground fault render the system unbalanced, thereby requiring special treatment through the application of symmetrical components. In this concept, the unbalanced system will be broken down into three balanced subsystems: namely, the positive-sequence system, the negative-sequence system, and the zero-sequence system as shown below in Fig. 8.1a, b, c, where

$$E_a = E_{a_1} + E_{a_2} + E_{a_0}$$
$$E_b = E_{b_1} + E_{b_2} + E_{b_0}$$
$$E_c = E_{c_1} + E_{c_2} + E_{c_0} \tag{8.1}$$

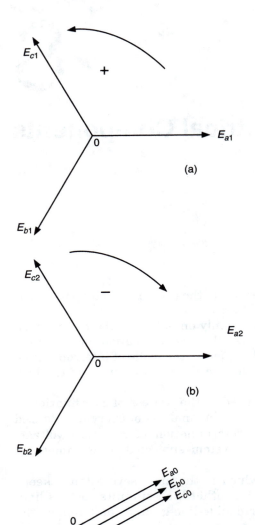

Figure 8.1 (*a*) Positive-, (*b*) negative-, and (*c*) zero-sequence systems.

let

$$\bar{a} = 1\angle120$$

$$\bar{a}^2 = 1\angle240$$

$$\bar{a}^3 = 1\angle360 = 1\angle0 \qquad (8.2)$$

Therefore, using Fig. 8.1*a, b, c,* and Eqs. (8.1) and (8.2), gives:

$$\begin{bmatrix} E_a \\ E_b \\ E_c \end{bmatrix} = \frac{1}{3} \begin{bmatrix} 1 & 1 & 1 \\ 1 & a^2 & a \\ 1 & a & a^2 \end{bmatrix} \begin{bmatrix} E_{a_0} \\ E_{a_1} \\ E_{a_2} \end{bmatrix} \tag{8.3}$$

or

$$\begin{bmatrix} E \\ a \\ b \\ c \end{bmatrix} = [T] \begin{bmatrix} E \\ a_0 \\ a_1 \\ a_2 \end{bmatrix} \tag{8.4}$$

where

$$T = \begin{bmatrix} 1 & 1 & 1 \\ 1 & a^2 & a \\ 1 & a & a^2 \end{bmatrix}$$

or

$$\begin{bmatrix} E \\ a_0 \\ a_1 \\ a_2 \end{bmatrix} = [T]^{-1} \begin{bmatrix} E \\ a \\ b \\ c \end{bmatrix} \tag{8.5}$$

Therefore,

$$\begin{bmatrix} E \\ a_0 \\ a_1 \\ a_2 \end{bmatrix} = \frac{1}{3} \begin{bmatrix} 1 & 1 & 1 \\ 1 & a & a^2 \\ 1 & a^2 & a \end{bmatrix} \begin{bmatrix} E \\ a \\ b \\ c \end{bmatrix} \tag{8.6}$$

Similarly, for sequence currents and phase currents we write:

$$\begin{bmatrix} I_{a_0} \\ I_{a_1} \\ I_{a_2} \end{bmatrix} = \frac{1}{3} \begin{bmatrix} 1 & 1 & 1 \\ 1 & a & a^2 \\ 1 & a^2 & a \end{bmatrix} \begin{bmatrix} E_a \\ E_b \\ E_c \end{bmatrix} \tag{8.7}$$

And, of course,

$$I_a + I_b + I_c + I_n = 0 \tag{8.8}$$

where I_n is the current in the neutral line in a Y system.

8.2 Power Calculations

The complex power vector $\overline{W}$ usually is written as follows:

$$\overline{W} = \overline{E}\,\overline{I}^*$$

since

$$\overline{V} = \overline{Z}\,\overline{I}$$

$$\overline{I} = \frac{\overline{E}}{\overline{Z}} = \overline{E}\ \overline{Y} \tag{8.9}$$

Therefore,

$$\overline{W} = |I|^2\ \overline{Z} = |E|^2\ \overline{Y}$$

$$= (P \pm jQ) \text{ in VA} \tag{8.10}$$

where P = the real power in watts
  Q = the reactive volt-ampere in VARs
  * indicates the conjugate of a vector

Also,

$$\overline{W} = \overline{E}_a\ \overline{I}_a^* + \overline{E}_b\ \overline{I}_b^* + \overline{E}_c\ \overline{I}_c \tag{8.11}$$

Using Eqs. (8.3), (8.4), and (8.11), this becomes:

$$\overline{W} = 3\ \overline{E}_0\ \overline{I}_0 + 3\ \overline{E}_1\ \overline{I}_1 + 3\ \overline{E}_2\ \overline{I}_2^* \tag{8.12}$$

8.3 Transformation of $\Delta - Y$

Frequently, in three-phase power systems, configurations of Δ and Y systems are interconnected. To simplify computations, it is easier to unify the overall system into a totally Δ and Y system.

Given the Δ load with three unbalanced impedances Z_1, Z_2, and Z_3 in Fig. 8.2, the required elements of the equivalent Y computed on the basis that the impedance at any two terminals is invariant are given as:

$$Z_A = \frac{Z_1 + Z_3}{Z_1 + Z_2 + Z_3}$$

$$Z_B = \frac{Z_1 + Z_2}{Z_1 + Z_2 + Z_3}$$

$$Z_C = \frac{Z_2 + Z_3}{Z_1 + Z_2 + Z_3} \tag{8.13}$$

Next, given the Y load with three unbalanced impedances Z_A, Z_B, and Z_C, the equivalent Δ elements computed on the basis that the total impedance at any two terminals is invariant are given as:

$$Z_1 = \frac{Z_A Z_B + Z_A Z_C + Z_B Z_C}{Z_C}$$

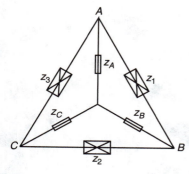

Figure 8.2 Transformation from Δ to Y system.

$$Z_2 = \frac{Z_A Z_B + Z_A Z_C + Z_B Z_C}{Z_A}$$

$$Z_3 = \frac{Z_A Z_B + Z_A Z_C + Z_B Z_C}{Z_B} \qquad (8.14)$$

It is timely to list the angular phase shift from the Δ side to the Y side of transformer, namely,

$$je_{A_1} = e_{a_1}$$
$$jI_{A_1} = I_{a_1}$$

and

$$je_{A_2} = e_{a_2}$$
$$jI_{A_2} = I_{a_2}$$

where the A subscript refers to the Δ side, and the a subscript refers to the Y side.

8.4 Unbalancing Due to Different Line Impedances

In Fig. 8.3, given that

$$Z_{aa'} \neq Z_{bb'} \neq Z_{cc'}$$

this will render the supply voltage drops:

$$E_{aa'} \neq E_{bb'} \neq E_{cc'}$$

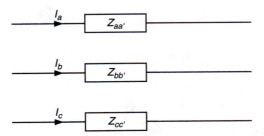

Figure 8.3 Unbalanced three-phase line impedances.

Therefore, this unbalanced picture could be expressed by:

$$\begin{bmatrix} E_{aa'} \\ E_{bb'} \\ E_{cc'} \end{bmatrix} = \begin{bmatrix} Z_{aa'} & 0 & 0 \\ 0 & Z_{bb'} & 0 \\ 0 & 0 & Z_{cc'} \end{bmatrix} \begin{bmatrix} I_a \\ I_b \\ I_c \end{bmatrix} \qquad (8.15)$$

where

$$E_{aa'} = E_{aa'_0} + E_{aa'_1} + E_{aa'_2}$$
$$E_{bb'} = E_{bb'_0} + E_{bb'_1} + E_{bb'_2}$$
$$E_{cc'} = E_{cc'_0} + E_{cc'_1} + E_{cc'_2} \qquad (8.16)$$

Also, I_a, I_b, and I_c are expressed similarly in terms of their zero-, positive-, and negative-sequence components. Then, from Eq. (8.15), we write:

$$E_{aa'} = I_a Z_{aa'}$$
$$E_{bb'} = I_b Z_{bb'}$$
$$E_{cc'} = I_c Z_{cc'}$$

The symmetrical sequence is:

$$E_{aa'_1} = I_{a_1} Z_{aa'}$$
$$E_{aa'_2} = I_{a_2} Z_{aa'}$$
$$E_{aa'_0} = I_{a_0} Z_{aa'}$$

8.5 Modes of Systems Interconnections

The well-known kinds of asymmetrical faults in terms of their severity are:

- three-phase short circuits
- double-line-ground short circuits
- single-line ground short circuits

The three sequence networks at the location of a fault are shown in Fig. 8.4, whereby

$$e_{a_1} = E_a - I_a Z_1$$
$$e_{a_2} = -I_{a_2} Z_2$$
$$e_{a_0} = -I_{a_0} Z_0 \tag{8.17}$$

At the location of the fault, with $E_a = E_f$ as the prefault voltage, we will discuss the symmetrical faults.

Three-phase short circuit. The system will remain symmetrical involving the positive-sequence system network and its steady-state emf E_a, where

$$e_{a_1} = E_a - I_{a_1} Z_1 \tag{8.18}$$

Double-line-ground short circuit. See Fig. 8.5. Initial conditions are at

$$t = 0^+$$
$$E_c = E_b = 0$$
$$I_a = 0 \tag{8.19}$$

Using Eqs. (8.1) through (8.8), the mode of sequence-networks interconnection is described by:

$$e_{a_1} = e_{a_2} = e_{a_0} \tag{8.20}$$

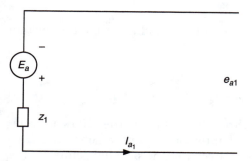

e_{a1}

Figure 8.4 Positive-sequence network.

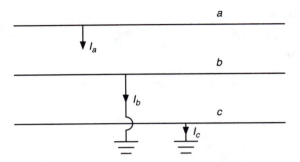

Figure 8.5 Double-line-to-ground fault.

$$I_{a_1} = \cfrac{E_i}{Z_1 + \cfrac{Z_2 Z_0}{Z_2 + Z_0}} \tag{8.21}$$

$$I_b + I_c = I_n \tag{8.22}$$

where Z_0 is the total zero-sequence impedance, and is given by:

$$Z_0 = 3Z_n + Z_{N_0} \tag{8.23}$$

where Z_n = the actual impedance inserted in the neutral line
Z_{N_0} = the zero-sequence impedance of the Y system
E_f = the prefault voltage at the location of the fault and is usually equal to 1.0 p.u. indicated as follows:

$$\begin{bmatrix} e_{a_0} \\ e_{a_1} \\ e_{a_2} \end{bmatrix} = \begin{bmatrix} 0 \\ E_f \\ 0 \end{bmatrix} - \begin{bmatrix} Z_0 & 0 & 0 \\ 0 & Z_1 & 0 \\ 0 & 0 & Z_2 \end{bmatrix} \begin{bmatrix} I_{a_0} \\ I_{a_1} \\ I_{a_2} \end{bmatrix} \tag{8.24}$$

Equation (8.18) indicates that for a double-line-ground short circuit, the three sequence networks are to be connected in parallel, which is shown in Fig. 8.6.

Line-line short circuit.   See Fig. 8.7. Initial conditions for the system shown in Fig. 8.6 are:

$$I_b = -I_c$$

$$e_b = e_c \tag{8.25}$$

Using Eq. (8.25) in Eqs. (8.1) through (8.8) gives the following relationships describing the mode of interconnections among the three sequence networks of the system involving the parallel connection of

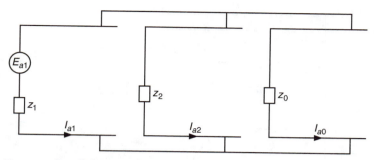

Figure 8.6 Parallel connection of sequence networks.

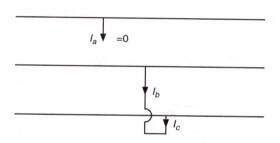

Figure 8.7 Line-line short circuit.

the positive- and negative-sequence networks and the isolation of the zero-sequence network. This is shown in Fig. 8.8.

$$e_{a_1} = e_{a_2}$$

$$I_{a_1} = \frac{E_f}{Z_1 + Z_2} \tag{8.26}$$

Line-ground short circuit. See Fig. 8.9. Initial conditions for this system are:

$$I_b = I_c = 0$$

$$e_a = 0 \tag{8.27}$$

Using the initial conditions of Eq. (8.27) in Eqs. (8.1) through (8.8) will produce the following relationships describing the mode of sequence network interconnections, namely,

$$I_{a_1} = I_{a_2} = I_{a_0} = \frac{E_f}{Z_1 + Z_2 + Z_0} \tag{8.28}$$

Equation (8.28) indicates that the mode of interconnections of the three sequence networks must be in series, as shown in Fig. 8.10.

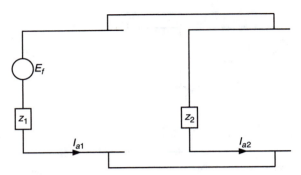

Figure 8.8 Parallel connection of positive- and negative-sequence networks.

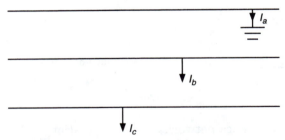

Figure 8.9 Line-ground short circuit.

Other kinds of faults which may not create serious damaging consequences are:

- one open conductor
- two open conductors

One open conductor. See Fig. 8.11. The initial conditions for this type of system are:

$$I_a = 0$$
$$V_{bb'} = V_{cc'} = 0 \tag{8.29}$$

Eq. (8.29) represents similar initial conditions for the case of a double-line-ground short circuit. Therefore, the mode of sequence network interconnections will be that the three sequence networks will be connected in parallel.

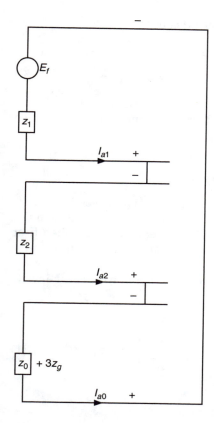

Figure 8.10 Series connection of sequence networks.

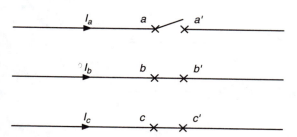

Figure 8.11 One open-conductor system.

Two open conductors.   See Fig. 8.12. Initial conditions are:

$$I_a = I_b = 0$$

$$V_{cc'} = 0 \qquad\qquad (8.30)$$

Initial conditions set by Eq. (8.30) are similar to the case of a single-line-ground short circuit, whereby the three sequence networks will be connected in series.

8.6   Bus Impedances in Fault Calculations

In multibus-bars power system, the interconnection of sequence networks at any bus-bar is based on bus-bars linkage with respect to the kind of fault. In power networks, usually it is easier to establish first the bus-admittance matrix and, by inversion, the bus-impedance matrix can be obtained.

For a system of four nodes, the bus-admittance matrix can be written as follows:

$$Y_{bus} = \begin{bmatrix} Y_{11} & Y_{12} & Y_{13} & Y_{14} \\ Y_{21} & Y_{22} & Y_{23} & Y_{24} \\ Y_{31} & Y_{32} & Y_{33} & Y_{34} \\ Y_{41} & Y_{42} & Y_{43} & Y_{44} \end{bmatrix} \qquad (8.31)$$

$$[Y_{bus}] = [Z_{bus}]^{-1}$$

$$= \frac{Adj\,[Z_{bus}]}{|Z|} \qquad (8.32)$$

where $ADj[Z_{bus}]$ = the Adjoint of $[Z_{bus}]$
$|Z|$ = the determinant of $[Z_{bus}]$

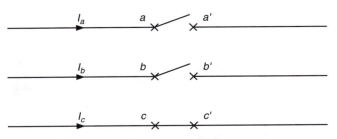

Figure 8.12 Two open-conductors system.

Usually the driving-point admittance at any bus-bar is given by:

$$Y_{nn} = \sum y_{nn} \qquad (8.33)$$

$$Y_{nm} = -\sum y_{nm} \qquad (8.34)$$

The mode of interconnection among the three sequence networks using elements of the bus-impedance matrix is the same as that in using circuit concepts. Figure 8.13 shows a case involving a double-line-ground short circuit.

8.7   Problems

8.1   Using Eqs. (8.6) and (8.7), derive the result shown in Eq. (8.11) for the total complex power vector under assymmetrical faults.

8.2   Prove mathematically using circuit theory rules Eq. (8.13), transforming a Δ load to the equivalent Y.

8.3   Repeat Prob. 8.2 with respect to Eq. (8.14) for the transformation of a Y load to the equivalent Δ.

8.4   Given that:

$$[T] \begin{bmatrix} E_{aa'_0} \\ E_{bb'_1} \\ E_{cc'_2} \end{bmatrix} = \begin{bmatrix} E_{aa'} \\ E_{bb'} \\ E_{cc'} \end{bmatrix}$$

then, using Eq. (8.15), derive relationships for $E_{aa'_1}$, $E_{aa'_2}$, and $E_{aa'_0}$ in terms of $Z_{aa'}$, $Z_{bb'}$, and $Z_{cc'}$, as well as I_{a_1}, I_{a_2}, and I_{a_0}.

8.5   In reference to the occurrence of double-line-ground fault at a three-phase power system, prove the criterion indicated in Eq. (8.18).

8.6   In reference to the occurrence of line-line short circuit on a power system, prove the criterion of network sequence interconnection given by Eq. (8.22).

8.7   In reference to the occurrence of line-ground short circuit on a power system, prove the criterion of network sequence interconnection given by Eq. (8.24).

8.8   Equation (8.27) gives general representation for the bus-admittance matrix for a four-bus-bar system. Using Eq. (8.28), obtain the equivalent bus-impedance matrix.

8.9   A matrix expressed in the principle coordinate system is known to have only diagonal elements. Such a matrix can be obtained by solving the determinant equation for λ_s:

$$\begin{vmatrix} Y_{11} - \lambda_1 & Y_{12} & Y_{13} & \cdots \\ Y_{21} & Y_{22} - \lambda_2 & Y_{23} & \cdots \\ \vdots & & & \\ \lambda_{n1} & \cdots & Y_{nn} - \lambda_n & \end{vmatrix}$$

The new matrix in the principle coordinates is given by:

$$\begin{bmatrix} \lambda_1 & & & \text{zeros} \\ & \lambda_2 & & \\ & & \lambda_3 & \\ \text{zeros} & & & \ddots \\ & & & & \lambda_n \end{bmatrix}$$

Therefore, given the following Y matrix, extract from the corresponding matrix in the principle coordinate system:

$$Y = \begin{vmatrix} -3.33 & 1.11 & 0 & 1.11 \\ 1.11 & -5.72 & 1.11 & 3.33 \\ 0 & 1.11 & -3.88 & 1.11 \\ 1.11 & 3.33 & 1.11 & -6.9 \end{vmatrix}$$

8.10 Using the bus-admittance matrix in the principle coordinate system obtained in Prob. 8.9, secure the corresponding bus-impedance matrix.

8.8 References

1. Denno, K., *Engineering Economics of Alternative Energy Sources,* CRC Press, 1992.
2. Nilson, James, W., *Electric Circuits,* 3d edition, Addison Wesley Publishing Co., 1990.
3. Stevenson, W. D., *Elements of Power System Analysis,* 4th edition, McGraw-Hill Book Company, 1982.

Relays for Asymmetrical Short Circuits

9.1 Introduction

In this chapter, systematic discussion and analysis of polarity power relays and distance relays will be presented with direct connection to the subject matter of symmetrical components application on protection from short circuits.

A power polarity relay reacts mainly to the change of sign or direction of power flow at its terminals and, at the same time, blocks a fault current that occurs outside a predefined protection zone.

Power polarity relays are classified according to their characteristics and mode of applications, namely: the volt-ampere magnitude to which the relay reacts, such as the sinusoidal mode with zero-phase angle and other specified phase angles; the kind of asymmetrical short circuit; the relay intended to protect against; the mode of internal design that may include the induction type, the polarized type, the contactless type, etc.; the number of phases to which the polarity power relay (PPR) is connected; the number of elements contained in the PPR; the phase or sequence of the volt-ampere subjected at the PPR terminals, whether the conventional volt-ampere or the symmetrical components counterpart; the state of invariant volt-ampere or switched volt-ampere; and, finally, the starting base with respect to reacting coils and the selection of fault phase.

In high-tension systems, the PPR must act when power flow is in the range of 2 ~ 3 times the normal limit; the effect of the volt-ampere in the undamaged phases must be neutralized and, while the dead zone is to be at a minimum for three-phase short-circuit fault, it is preferred that it be eliminated completely with respect to other kinds of short circuits.

Volt-amperes operating at the relay terminals are reflected in terms of the resulting electromagnetic torque $T_{e,m}$, which is proportional to the product of the magnetic induction vector $\overline{B}$ and the field intensity vector $\overline{H}$. Product multiplication of $\overline{B}$ and $\overline{H}$ may be in terms of the dot product $\overline{B} \cdot \overline{H}$ or the cross multiplication $\overline{B} \times \overline{H}$. It is important to note that for a ferromagnetic relay core, the B field is a nonlinear function of the H field.

Current research efforts point to increasing interest in solid-state relays, but somehow the process is slow in replacing the already operational electromechanical relays.

9.2 Asymmetrical Sequence Power at Fault

During steady-state operation of a power system, positive-sequence components of voltage and current are the only entities present. However, at the occurrence of a short circuit, the negative- and zero-sequence powers flow into the fault due to the converted positive-sequence power existing before the fault in the absence of load.

In the presence of load, prefault positive-sequence power and volt-amperes will still represent positive-sequence-braking torque.

Therefore,

$$\overline{W}_1 = -(\overline{W}_2 + \overline{W}_0) \tag{9.1}$$

or

$$\overline{V}_1 \, \overline{I}_1^x = -[\overline{V}_2 \overline{I}_2^x + \overline{V}_0 \overline{I}_0^x] \tag{9.2}$$

x denotes conjugate where, in the case of double-line-ground short circuit,

$$\overline{V}_1 = -(\overline{V}_2 + \overline{V}_0) \text{ and}$$
$$\overline{I}_1 = \overline{I}_2 = \overline{I}_0 \tag{9.3}$$

where the three sequence networks are connected in parallel. Then, for the case of line-line short circuit,

$$\overline{V}_1 = \overline{V}_2 \text{ and}$$
$$\overline{I}_1 = \overline{I}_2 \tag{9.4}$$

where the negative- and positive-sequence networks are in parallel, while the zero-sequence network is isolated.

And, for a line-ground short circuit,

$$\overline{V}_1 = \overline{V}_2 = \overline{V}_0 \tag{9.5}$$

and

$$\overline{I}_1 = -(\overline{I}_2 + \overline{I}_0) \tag{9.6}$$

where the three sequence networks are connected in series.

Application of phase voltages at relay terminals implies the presence of V_0 (the zero-sequence voltage), while the application of line-line voltages will exclude the zero-sequence components.

Therefore, under the absence of zero-sequence volt-ampere, the operating power at the polarity power relay terminals is proportional to $(\overline{W}_1 + \overline{W}_2)$, that is,

$$W_{op} < R_e[\overline{W}_1 + \overline{W}_2]$$
$$< R_e[\overline{V}_1 \ \overline{I}_1^x + \overline{V}_2 \ \overline{I}_2^x] \tag{9.7}$$

where $<$ denotes proportional

R_e denotes the real part of any complex quantity

Then, for the presence of zero-sequence volt-ampere,

$$W_{op} < R_e \left[\overline{W}_1 + \overline{W}_2 + \overline{W}_0 \right]$$
$$< R_e \left[\overline{V}_1 \ \overline{I}_1^x + \overline{V}_2 \ \overline{I}_2^x + \overline{V}_0 \ \overline{I}_0^x \right] \tag{9.8}$$

Of course, W_{op} is, in effect, the operating electromagnetic torque.

Calculation of W_{op} for the PPR has to proceed toward the optimum limit, which involves computation of the complex terms on the right-hand side of Eq. (9.8). Under a no-load situation, the dominant part of the positive-sequence power will be converted to a combination of negative and zero sequences, or merely to only negative-sequence power for a line-line short circuit.

9.3 Criteria for PPR Relay Selection

Under the occurrence of a fault on a three-phase power system, computation has to be carried out for all volt-ampere values of the sequence-symmetrical components, taking into consideration the mode of relay connection to the three-phase system. For the case of line-neutral voltage connection which comes into the picture for a double-line-ground short circuit and a single-line-ground short circuit, all positive-, negative-, and zero-sequence voltages and currents must be calculated, while, in the case of a line-line short circuit, only the positive- and negative-sequence volt-amperes ought to be calculated. Consequently, complex power vector $\overline{W}_1, \overline{W}_2$, and $\overline{W}_0$ will be computed and a ratio of $\overline{W}_2/\overline{W}_1$ will be set, and angular phase shift for $\overline{W}_1$

(say, Φ_1) and $\overline{W}_2$ (say, Φ_2) will be identified concurrently with the setting of specific internal phase angles for the relay (which are internal design characteristics). Those relay internal angles usually are special angles characterized as 0°, 15°, 30°, 45°, 60°, 90°, and 120°.

The following main rules or criteria will serve as a guide to select the most optimal relay to provide compatible protective response for the PPR to any prescribed short circuit:

1. Magnitudes of operating torques generated by the positive-, negative-, and zero-sequence complex power vectors are preferred to be maximum.

2. In the case of line-line short circuit, operating torque generated by the negative-sequence complex power is preferred to be as large as possible with respect to that generated by the positive-sequence complex power (which may be aided with prefault complex positive-sequence power in the case of load).

3. Maximum ratio of $|\overline{W}_2|/|\overline{W}_1|$ should be weighted with respect to the relay internal angle, as well as with respect to a particular value of Φ_2 and Φ_1, and are preferred to be maximum.

4. Under loading conditions, the prefault current components cause a working torque at one end of the protected zone and a braking torque at the other end. The preference is to have a contribution to the negative-sequence torque as a working torque.

5. Effect of the braking torque which comes from the positive-sequence components is preferred to be minimum (i.e., a contribution to the positive-sequence torque).

Case Study I. A three-phase Y-connected AC generator having $X_1 = j$ 0.25 p.u., $X_2 = j$ 0.25 p.u., and $X_0 = j$ 0.08 p.u. is subjected to a line-line short circuit and operated at rated voltage and disconnected from the system. The following is given:

$$e_{a_1} = 0.5\angle 0°\text{p.u.}, \quad I_{a_1} = 2.0\angle -90°\text{p.u.}$$
$$e_{a_2} = 0.5\angle 0°\text{p.u.}, \quad I_{a_1} = 2.0\angle 90°\text{p.u.}$$

Therefore,

$$\Phi_1 = 90°, \quad \Phi_2 = -90°$$

Using the general guidelines mentioned earlier, it is required to present optimal choices and the optimum selection of polarity power relay to respond adequately to the line-line short circuit.

Power polarity relay is usually identified by mode angles and internal phase angles. Angular modes are special angles which may be the same as the internal design angle or different.

In line-line short circuit, zero-sequence network is isolated and, hence, there is no connection to the combination of positive- and negative-sequence networks which are to be combined in parallel for this kind of fault.

Therefore, T_{oper}, is given by:

$$T_{oper} < R_e \left[\overline{W}_1 + \overline{W}_2 \right] \tag{9.9}$$

where

$$\left| W_1 \right| = K R_e \left[e_{a_1} I_{a_1}^x \right] = T_1 \text{ newton-meter} \tag{9.10}$$

$$\left| W_2 \right| = K R_e \left[e_{a_2} I_{a_2}^x \right] = T_2 \text{ newton-meter} \tag{9.11}$$

where x denotes conjugate.

Angular modes (θ_m) and internal phase shift angles (θ_d) are selected randomly by pair as shown in Table 9.1:

Calculating T_{oper} for the first selection of any θ_m and θ_d is as follows:

$$T = T_1 + T_2$$
$$= 3\sqrt{3}k \, [e_{a_1} i_{a_1} cos(\theta_1 - \theta_m + \theta_d)$$
$$+ e_{a_2} i_{a_2} cos(\theta_2 - \theta_m + \theta_d)] \tag{9.12}$$

where

$$T_1 = 3\sqrt{3}k e_{a_1} i_{a_1} \, cos(\theta_1 - \theta_m + \theta_d) \tag{9.13}$$

$$T_1 = 3\sqrt{3}k e_{a_2} i_{a_2} \, cos(\theta_2 - \theta_m + \theta_d) \tag{9.14}$$

Calculation of T_1, T_2, and T_{oper} has been repeated for every pair of θ_m and θ, resulting in the values shown in Table 9.2.

TABLE 9.1

θ_m^0	30	30	60	60	90	90	90	120	120	120
θ_d^0	0	30	0	30	0	30	45	0	30	45
Selection no.	1	2	3	4	5	6	7	8	9	10

TABLE 9.2

T_1	2.6k	0	2.6k	1.5k	5.2k	4.5k	3.68k	7.79k	9k	8.7k
T_2	2.6k	4.5k	2.6k	3.0k	5.2k	4.5k	3.68k	7.79k	7.79k	2.33k
T_{op}	5.2k	4.5k	5.2k	4.5k	10.4k	9.0k	7.36k	15.58k	16.79k	11.03k
Selection	1	2	3	4	5	6	7	8	9	10

By inspecting Table 9.2 using the guidelines mentioned in this section, the following optimal selections could be identified: of selection numbers 5, 6, 7, 8, and 9, the most optimum choice is selection number 8, where $T_2 = T_1$. Selection number 9 is not the most optimum because $T_2 < T_1$.

9.4   Distance Relays

Performance of distance relays is in response to changes in voltages and currents at their terminals at a location in the power system set by the impedance, phase angles, and effects of other interconnected parts of the system. Action of distance relay is reflected in providing protection for a preselected zone and no action for anything outside that zone. Protection includes the role of starting where the distance relay responds at the start of a short circuit and, in the process of selection, distance relay responds to its protective role in a system where various operations are supposed to be provided. A well-known example of a distance relay is the ohm relay, whose response is to differentiate the ohmic value as well as the phase angle at the location of a short circuit.

Working conditions at the relay terminals depends upon location of its vector impedance locus, specifically under an overload or a short circuit. Under short circuit, location of the impedance vector will be associated with a much higher lag of phase angle than that on an acceptable overload.

Time-lag characteristics of distance relays are of two modes, namely, the step-definite and the step-truncated, ramp-definite modes expressed mathematically as follows:

For the step-definite mode:

$$t = A_1 U_{-1}(t - T_1) + A_2 U_{-1}(t - T_2) + A_3 U_{-1}(t - T_3) + \cdots + A_n U_{-1}(t - T_n)$$

$$(9.15)$$

Then, for the step-truncated, ramp-definite mode:

$$t = A_1 U_{-1}(t - T_1) + A_2 U_{-1}(t - T_2) + A_3 U_{-1}(t - T_3) + \cdots$$
$$+ A_{n-1} U_{-1}(t - T_{n-1}) - A_n t U_{-1}(t - T_n) \qquad (9.16)$$

Classes of distance relays are usually characterized as *directional* and *nondirectional*. Directional distance relays are identified by the character of their impedance or admittance vector locus. This locus could be in the form of a displaced circle not passing into the origin of the two-dimensional plane or with its circumference passing through the origin. Nondirectional relays may have their center coincide with the origin for the impedance or admittance type. Also directional or nondirectional distance relays may have an elliptical impedance or admittance locus, performing the role of blocking relay. Tripping action of distance relays is, in general, in response to symmetrical or asymmetrical short circuits.

Clarification for the selected region of operation for a distance relay represented by its impedance or admittance locus could be done by the process of conformal mapping, i.e., from the x-y plane to the u-v plane. The process of plane transformation is intended to represent the locus in a new plane more compatible with the physical reality for the protected zone and for a particular power system.

Well-known configurations that usually require transformation through conformal mapping are shown in Figs. 9.1, 9.2, 9.3, and 9.4.

Figure 9.1 transforms a confined, protected region of the interior of a circle in the x-y plane to an unlimited area outside a circular characteristic in the u-v plane.

Figure 9.2 transforms a circular characteristic for a region to be protected in the x-y plane, passing in the origin to an unlimited area to the right of the plane $u \geq B'$ in the u-v plane.

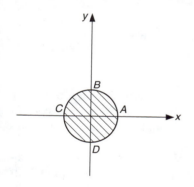

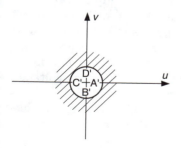

Figure 9.1 Mapping the interior of circle into the exterior of circle by the function $W = \frac{1}{z}$.*

---

* Refer to Ref. 2 at the end of this chapter.

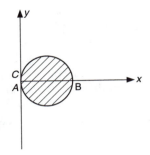

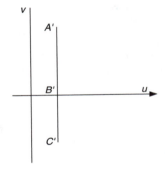

Figure 9.2 Mapping of side circle into strip by the function $W = \frac{1}{2}.$*

Figure 9.3 transforms the protected region having its impedance or admittance locus covering the entire upper-half plane in the x-y system to a circular bounded region whose center coincides at the origin in the u-v plane.

Figure 9.4 transforms a locus area for the Z or Y vector for a protected region, unlimitedly bounding a half circle in the upper-half plane of the x-y system to the entire upper-half plane in the u-v plane.

Usually, other functions for transformation from the x-y plane to the u-v plane can be found in many reference books on complex variables.

Admittance at distance relay terminals reflects ratio of current applied to the source emf. Admittance loci are usually in the form of concentric circles of constant emf with their radial lines representing constant-phase angles. Distance relays for line-line short-circuit faults are connected in a mode of current differentials and line-line voltages, while those for line-ground fault are connected in a mode of phase voltage and compensated current. Distance relays intended for double-line-ground fault are to be connected in a mode of two line-neutral voltages and compensated currents.

---

* Refer to Ref. 2 at the end of this chapter.

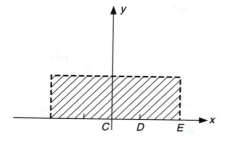

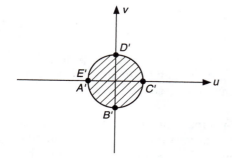

Figure 9.3 Mapping of semi-infinite strip into the interior of circle by the function $W = i - z/i + z.$*

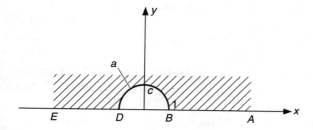

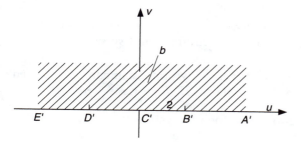

Figure 9.4 Mapping of area (a) into (b) by the function $W = z + \frac{1}{2}.$*

---

* Refer to Ref. 2 at the end of this chapter.

For line-line fault, with relays in lines AB, BC, and CA, calculations have to be carried out to compute Y_{AB}, Y_{BC}, and Y_{CA} in terms of symmetrical sequence currents with respect to prefault emf at the source. Similarly, for a line-ground fault, Y_A, Y_B, and Y_C have to be calculated in terms of symmetrical sequence currents and prefault emf.

The operating characteristic of a distance relay will be displaced in the complex admittance plane by the positive-sequence impedance of the source, with the source emfs representing the voltage components applied at the relay terminals; then the new admittance locus will be the inversion of a displaced impedance locus by the incremental positive-sequence impedance in question.

Following is an outline for the calculation of protection zone by distance relays:

1. Computation of admittance at relay terminals with respect to line-line fault, single-line ground fault, or a double-line-ground fault can be obtained from calculated symmetrical sequence values of relevant voltages and currents.

2. Since voltages applied at the terminals of distance relay are the source emfs and their phase angles, admittances will be calculated at the relay terminals first when the receiving-end emf is zero and then again when the emfs at the receiving and sending ends are equal.

3. The difference between the admittance with the receiving end short-circuited and the admittance with the emf at the sending end equal to that at the receiving end will identify the radius of the admittance circle which is valid when $E_s = E_R$.

4. Specification of the zone impedance to be protected by the distance relay and the sending end emf with its phase shift will establish the Z circle.

5. Inversion of the Z circle into the Y plane will result in the intersection of the inverted Z circle with the Y circle at the relay terminals established in step 3.

6. Angular zone of distance relay coverage can be determined by the span between two arrays passing through the two points of intersection, as indicated in step 5.

7. A wider angular scope of distance relay protection can be obtained by drawing another set of Y circles for $E_s/E_R > 1$.

Case Study II.[*] In the two-source power system shown in Fig. 9.5, establish the angular phase coverage for a distance relay located at position R, with the predesigned protection of 50 Ω. The fault is a line-line short circuit between lines C and A at F.

---

[*] Refer to Ref. 1 at the end of this chapter.

Consider $Z_1 = Z_2 = Z_0$ for the system. The equivalent circuit is shown in Fig. 9.6. Relevant circuit-loop equations are:

$$E_s = i_1(35\angle 45°) + (i_1 - i_3)(35\angle 45°)$$
$$E_r = i_2(45\angle 45°) + (i_2 - i_3)(45\angle 45°)$$
$$0 = (i_3 - i_1)(35\angle 45°) + (i_2 + i_3)(45\angle 45°) \tag{9.17}$$

Solving from Eq. (9.17), the ratio of i_1/E_s and i_2/E_s when $E_r = 0$, identified as i_1/E_s and i_2/E_s, is given by:

$$\frac{I_1}{E_s} = 0.0205\angle -45° \tag{9.18}$$

$$\frac{I_2}{E_s} = -0.00803\angle -45° \tag{9.19}$$

Then, repeating to solve for i_1/E_s and i_2/E_s when $E_s = E_r$ is also identified as:

$$\frac{I_1'}{E_s} = 0.0143\angle -45° \tag{9.20}$$

$$\frac{I_2'}{E_s} = -0.0143\angle -45° \tag{9.21}$$

Figure 9.5 Two-source power system.

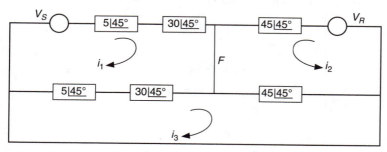

Figure 9.6 Equivalent circuit of Fig. 9.5 for line-line short circuit.

From symmetrical component sequences, we obtain:

$$Y_{CA} = \frac{I_1 - a^2 I_2}{E_s} \tag{9.22}$$

Y_{CA} at $E_r = 0$

$$= 0.0205\angle{-45°} + \angle{240°}(0.00803\angle{-45°})$$

$$= 0.0179\angle{-68°} \tag{9.23}$$

Y_{CA} at $E_r = E_s$

$$= 0.0143\angle{-45°} + \angle{240°}(0.0143\angle{-45°})$$

$$= 0.0143\angle{-105°} \tag{9.24}$$

$$Y_{CA} = Y_{CA}\big|_{E_r=0} + [Y_{CA}\big|_{E_s=E_r} - Y_{CA}\big|_{E_r=0}] \frac{E_r}{E_s}$$

$$= 0.0179\angle{-68°} + [0.0143\angle{-105°} - 0.0179\angle{-68°}] \frac{E_r}{E_s} \tag{9.25}$$

The admittance loci for Y_{CA} at $E_r/E_s = 1.0$ and 1.50 are shown in Fig. 9.7. Also, a plot for the Z locus, a circle of diameter 50 Ω with vector deviation of $5\angle{45°}$ Ω, is shown. Subsequently, inversion of the Z circle established a Y locus. Intersection of Y_A loci with the inversion of the Z circle identifies the angular span of phase coverage for the protected zone:

$$\text{For } E_r/E_s = 1, \quad 127° < \theta < 289°$$

$$\text{For } E_r/E_s = 1.50, \quad 128° < \theta < 292°$$

9.5 Problems

9.1 A balanced three-phase system is subjected to a line-ground short circuit, characterized by the following symmetrical components:

$$i_{a_1} = 4.21\angle{-90°} \text{ p.u.}$$

$$i_{a_2} = 2.17\angle{90°} \text{ p.u.}$$

$$i_{a_0} = 2.04\angle{90°} \text{ p.u.}$$

At the location of fault, the sequence impedances are:

$$Z_1 = Z_2$$

$$= j0.50 \text{ p.u.}$$

$$Z_0 = j0.05 \text{ p.u.}$$

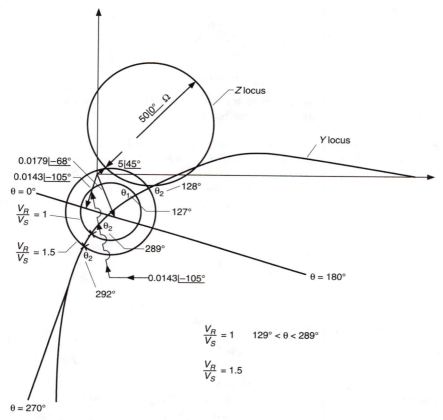

Figure 9.7 Loci of Y_{CA}.

Prefault p.u. phase voltage = 1.0 p.u. is required to carry out a process of torque calculations for the positive-, negative-, and zero-sequence components. Using the criteria of selectivity, identify the most optimal mode of adequate polarity power relay. Assume the system was operating at no-load before occurrence of short circuit.

9.2 A balanced three-phase system is subjected to a double-line-ground short circuit, characterized by the following symmetrical components:

$$i_{a_1} = -j3.90$$

$$i_{a_2} = j1.34$$

$$i_{a0} = j2.56$$

$$e_{a_1} = e_{a2} = e_{a_0} = j7.68 \text{ all in p.u.}$$

E_f: the prefault p.u. voltage = 1.0 p.u.

$$Z_1 = j0.188 \text{ p.u.}$$

$$Z_2 = j0.2 \text{ p.u.}$$

$$Z_0 = j0.104 \text{ p.u.}$$

Carry out a process of symmetrical torque calculations to determine the most optimal mode of adequate polarity power relay to be selected.

9.3 A balanced three-phase system is subjected to a line-line short circuit, characterized by the following symmetrical components:

$$i_{a_1} = -j1.667 \text{ p.u.}$$

$$i_{a_2} = j1.667 \text{ p.u.}$$

$$i_{a_0} = 0$$

$$Z_1 = j0.25 \text{ p.u.}$$

$$Z_2 = j0.35 \text{ p.u.}$$

$$Z_0 = j0.10 \text{ p.u.}$$

$$E_f = 1.0$$

Carry out a process of symmetrical torque calculations to determine the most optimal mode of polarity power relay to adequately provide protection.

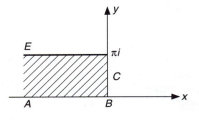

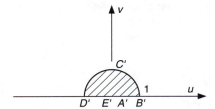

Figure 9.8 Problem 9.4.*

---

* Refer to Ref. 2 at the end of this chapter.

9.4 A distance relay has its admittance locus in the x-y plane as a semicircle in the upper-half plane. Identify the conformal transformation function for converting the semicircle in the x-y plane area in the u-v plane defined by $v = \pi j$ and $u = 0$.

9.5 In reference to Case Study II, identify the angular span of phase coverage when the fault at point F is a double-line-ground short circuit. The impedance for the zone to be protected by distance relay $= 80\ \Omega$.

9.6 Repeat Prob. 9.5 for the case of line-ground short circuit where the impedance of the zone to be protected is $40\ \Omega$.

9.7 Given that the impedance vector locus for a distance relay covering the entire region in the right-half plane $u \geq 0$. $-\infty \leq v \leq \infty$, identify with proof the required conformal mapping function of converting the described right-half plane to a circle whose center is the origin in the x-y plane.

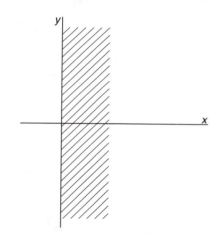

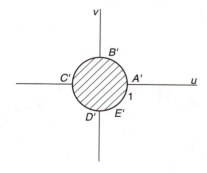

Figure 9.9 Problem 9.5.*

---

* Refer to Ref. 2 at the end of this chapter.

9.6   References

1. Atabekov, G. I., *The Relay Protection of High Voltage Networks,* Pergamon Press, London, Oxford, New York, and Paris, 1960.
2. Churchill, R. V., *Introduction to Complex Variables and Applications,* McGraw-Hill Book Co., 1948.
3. Denno, K., *High Voltage Engineering in Power Systems,* CRC Press, 1992.
4. *Elements of Power System Analysis,* 3d edition, McGraw-Hill Book Co., 1973.
5. Warrington, A. R., Van C., *Protective Relays, Their Theory and Practice,* volume 2, John Wiley & Sons, Inc., New York, 1969.

10

Power Electronics
in Protection System

In this chapter, the future mode of power generation, transmission, distribution, and their intrinsic structure of protection through the involvement of solid-state technology will be presented. The presentation in this chapter is based on the role of power electronics—specifically, silicon thyristor in the transmission of bulk electric power with prompt protection—and will be comprised of two articles published in the *EPRI Journal* (reprinted here by permission). EPRI is the acronym for Electric Power Research Institute.

10.1   Sealed in Silicon—the Power Electronics Revolution*

Development of power electronics—solid-state technology for the efficient handling of bulk electric power—stands at the point where integrated circuits did 30 years ago. Like their microelectronic counterparts, power electronics devices are expected to have a revolutionary effect, changing forever the way we generate, deliver, and use electricity.

It has been called the second electronics revolution—the application of advanced semiconductor technology to high-voltage, high-current tasks in utility networks, industrial process, and home appliances. The advantages of speed, reliability, and efficiency that integrated circuits

---

* This article was written by John Douglas, science writer. Technical background information was provided by Narain Hingorani and Harshad Mehta, Electrical Systems Division; Ralph Ferraro, Energy Management and Utilization Division; and Frank Goodman, Advanced Power Systems Division.

on silicon chips brought at the microwatt level to computers are now being used to control megawatt transmission networks through power electronics.

In the simplest terms, solid-state electronics devices, both large and small, gain advantages of speed and low energy loss by whisking current through alternate layers of silicon with different conducting properties. In integrated circuits, individual devices such as diodes and transistors are microscopic. In power electronics equipment, a single switching device called a *thyristor* is constructed from four silicon layers that may be 4 in (102 mm) in diameter for high-current applications. Sealed in a modular package, such a device can weigh 5 lb (2.3 kg). Equipment like a transmission line ac/dc converter may consist of hundreds of thyristors arranged in approximately 50-ft (15-m) stacks. Just as integrated circuits replaced vacuum tubes in radios, thyristors have already replaced expensive, inefficient mercury arc tubes in ac/dc converters and may someday be able to replace mechanical circuit breakers on distribution lines.

The full impact of power electronics devices is just beginning to be felt, for they are constantly finding new applications ranging from adjustable speed for motors to industrial heating controls. "Within about two decades," predicts Narain Hingorani, vice president, Electrical System Division, "all electricity will flow through several power semiconductor stages between generation and consumption."

Concern is rising, however, that the United States is falling behind overseas competitors in the development and use of power electronics. The pure, defect-free, high-resistivity, single-crystal silicon required for power semiconductors is not now commercially available from a domestic source. Several important areas of R&D related to the design and packaging of high-power devices are lagging because of insufficient research funds. And some technology already available is not being fully utilized because equipment manufacturers are often unaware of the advantages power electronics offers for lowering capital costs and improving productivity and energy efficiency.

Electric utilities probably have the most at stake in the international competition over power electronics. Within the utility industry itself, power semiconductor devices are revolutionizing control and conversion of bulk electricity. Greater use of power electronics in industry could also benefit utilities by improving load efficiency and increasing market penetration of new electrotechnologies.

For the last 12 years EPRI has sponsored pioneering research on advanced power semiconductor devices for utility application. This research has helped establish U.S. leadership in specific areas of power electronics technology and has led to a dramatic increase in the use of power semiconductors in some key transmission operations. Further progress, however, will require closer coordination of increasingly expensive research by producers of semiconductor materials, equipment manufacturers and users, government organizations, and the utility industry. EPRI is now helping to establish such a coordinate national research effort, which will be unprecedented in its scope and potential effect on a very fragmented area of high technology.

Research on power electronics can be roughly divided into three broad areas: semiconductor materials, devices and packaging, and applications. At EPRI the Electrical Systems Division is responsible for R&D on semiconductor materials, advanced devices, and applications related to high-voltage dc (HVDC) and other transmission and distribution areas. Applications of power semiconductor devices to industrial and commercial equipment and systems is handled by the Energy Management and Utilization Division; the Advanced Power Systems Division evaluates new materials for solar photovoltaic energy conversion and is developing power electronics controls to improve wind turbine generators and battery storage systems.

The challenge of materials. Materials research is fundamental because device options are limited by the properties of the semiconductors from which they are made, and materials costs are leveraged through production of highly value-added electronic systems. (In 1985 worldwide consumption of all electronic materials was $2.4 billion, which supported at $388 billion.) Because of this leverage, the rapid erosion of an early U.S. lead in materials technology has caused mounting concern. A national workshop sponsored by the Federation of Materials Societies recently stated, "The loss of a competitive electronic materials base inevitably leads to a loss in competitiveness in the manufacture of circuits and systems." So rapidly is technology changing, the group concluded "if the current trend continues, it can be anticipated that the United States will be a minor force in the world market in electronic materials and systems by the early 1990s."

This conclusion applies particularly to the special semiconductor materials needed for high-power electronics devices. The single-crystal silicon used in the manufacture of integrated circuits represents some of the purest material ever produced, but it still contains too many impurities to withstand the strain of high voltages and currents. Most of this quality material is produced in ingots with a diameter of 3 ~ 5 in (76 ~ 127 mm) drawn vertically from the surface of molten silicon in a quartz crucible. The problem with this Czochralski (CZ) method is that oxygen and other impurities from the crucible are incorporated into the silicon crystal structure.

Currently, all the silicon for power electronics is produced by a float zone (FZ) method, which does not involve a crucible. Instead, a rod of polycrystalline silicon is suspended in a radio-frequency field, which melts its lower portion and allows a single-crystal ingot to be drawn downward with very little contamination. The problem is that the FZ process for producing silicon ingots, which are sliced into wafers for device manufacture, is much more expensive than the CZ method. Also, the only commercial FZ facilities are outside the United States—one in West Germany and two in Japan. These manufacturers give preference to their domestic buyers of power semiconductor material, so the lead time for filling orders for the United States and other countries can be as long as 60 weeks.

EPRI has recently granted two manufacturing licenses and associated R&D contracts for producing power-quality silicon in an effort to lower costs and advance the silicon ingot technology available in the United States. Cybeq, a division of Siltec Corp., received a license and contract to explore the feasibility of improving the quality of silicon produced by the CZ method by surrounding the crucible with an intense magnetic field. The presence of a field greatly slows the migration of impurity ions through the silicon melt. If successful, this magnetic CZ process should reduce the cost of power-quality silicon by 15 ~ 20%, compared with using the conventional FZ method. At the same time Westinghouse Electronic Corp. will explore ways of improving the FZ process and making it less expensive. Total savings for the utility industry from these programs are expected to run $30 million a year by 1995.

Devices and packaging. The usefulness of solid-state electronics devices stems from their ability to shift the flow of electricity in response to very subtle changes inside their semiconductor material. Because of their advantages in handling high currents, thyristors have become the workhorse of power electronics. EPRI work has focused on reducing the cost of thyristors, increasing their versatility, and lowering the energy losses that occur in these important devices.

One important breakthrough occurred in 1983 with the first commercial demonstration of a light-fired thyristor developed by General Electric Co. and Westinghouse in cooperation with EPRI. Use of light greatly reduces the energy needed to turn on the thyristor and reduces overall system cost because less auxiliary equipment is necessary. During the successful EPRI-sponsored demonstration, a combination of electrically triggered and separate, light-fired thyristors were part of the HVDC module installed at the southern end of the Pacific Intertie line, the country's largest HVDC transmission system. Development of the light-fired thyristor is expected to have significant impact in the near future on lowering the cost and increasing the reliability of HVDC converter valves.

A nagging problem since thyristors were first developed has been the need to reverse line polarity before they can be turned off. One solution has been to develop a gate turn-off (GTO) thyristor by rearranging the device configuration so the gate is nearer the negative (cathode) end of the device. When a negative voltage is connected to the gate the thyristor turns off, enabling it to serve as a two-way power switch. Such a capability can considerably widen the applications in which thyristors are used, and EPRI is involved in both cost reduction research on GTOs and efforts to demonstrate their potential benefits.

Recently, a contract was issued to General Electric to use the most advanced 3-in-diam (77-mm) GTOs now commercially available as the basis for the power conditioning system in Southern California Edison Co.'s new battery energy storage facility.

The main disadvantage of GTOs is the large gate current needed to turn them off. Over the long term, a more promising technology appears to be the so-called MOS-controlled thyristor, or MCT. (MOS stands for Metal Oxide Semiconductor and refers to the arrangement of components in devices based on this technology. Many integrated circuits requiring very low energy loss, such as those in watches, are based on MOS technology.)

An MCT would consist essentially of an MOS integrated circuit created on the top surface of a high-power thyristor. In such a device a very large line current could be switched off by a much smaller gate current. In addition, MCTs have a turn-off time less than one-third that of GTOs. Considerable R&D still has to be conducted on MCTs, however, before their full potential can be reached. This work includes raising the voltage and current capacity, improving circuit density and device yield through better fabrication, and introducing new package concepts. General Electric recently received a license to manufacture a high-power MCT developed through EPRI funding.

The issue of device packaging arises frequently in discussions of research still needed on all kinds of power semiconductor devices. The basic reason is that present devices generally use a large tungsten or molybdenum contact plate to transfer current from a silicon wafer at the heart of a device to the rest of a high-power circuit. If a contact plate made of silicon could be used, the cost and weight of power thyristors would be greatly reduced. The challenge to using silicon contact plates is that they have more resistance and less ability to conduct heat than does metal. These and other research concerns are being addressed through EPRI contracts with General Electric and Powerex, Inc. (a power semiconductor joint venture company owned by General Electric, Westinghouse, and Mitsubishi).

"The revolution in power electronics is being driven by the need to lower costs, increase efficiencies, and conserve energy," explains Harshad Mehta, Electrical Systems Division project manager. "Power semiconductors are already proving their worth in a variety of utility applications, including HVDC transmission, static VAR compensators, subsynchronous resonance damping, and adjustable-speed motor drives. If EPRI's work on materials technology and advanced devices is successful, the use of power electronics in utility-related applications should expand dramatically over the next few years. This will be particularly true if our research on MCTs pays off because this technology has the potential for replacing most of the devices in use today."

Opportunities for application. Power electronics also has great potential to reduce costs and improve the efficiency of a wide variety of industrial and commercial technologies that depend on electricity. Devices already developed can cut the cost of adjustable-speed drives, power supplies, power-line conditioners, and other equipment by 50% or more. Properly integrated into such equipment, power semiconductors may contribute a 20% overall savings in capital costs of some equipment for industrial processes. Such savings could help lower manufacturing costs and increase industrial productivity, benefiting utilities by increasing load retention and efficiency, as well as by supporting load compatibility and growth.

Many industrial energy users and process equipment manufacturers, however, are unaware of the opportunities provided by power electronics. In addition, many of the power semiconductor devices now in use have been imported. Changes in the critical marker for motors with electronic adjustable-speed drive (ASD) are symptomatic. Since 1980, while total sales have risen steadily, the market for domestically produced ASD equipment has actually declined and the value of imports has more than tripled. United States industry is now considered to be somewhat behind other countries in the exploitation of power electronics.

To increase the awareness and understanding of power electronics technologies and issues, EPRI recently established the Power Electronics Applications R&D Center (PEAC) in Knoxville, Tennessee. The center will be a focal point for applications research of power electronics equipment and for transferring related technologies to the commercial sector. PEAC will be run under contract by the Tennessee Center for Research and Development, with staff and facilities support from the University of Tennessee, Knoxville, the Tennessee Valley Authority, Martin Marietta Energy Systems, Electrotek Concepts, Inc., and the state of Tennessee. More than two dozen other organizations, including several major corporations, have also pledged their support.

One of the first tasks to be undertaken at PEAC will be a thorough assessment of currently available power electronics technologies in

order to define specifications for new equipment, systems, and application. PEAC will relate these specifications to advances in component
and device technology and develop collaborative R&D programs with
industries to integrate these advances into commercial equipment.
Specific R&D projects will be launched to improve the use of power
electronics in selected applications, such as uninterruptible power supplies, active power-line conditioners, process control and energy management systems, and adjustable-speed drives. These projects will be
cofunded by the industries most likely to benefit.

"Our emphasis at PEAC will be two-way communication with the
technology users and the technology creators," comments Ralph Ferraro, program manager in the Energy Management and Utilization
Division. "The potential applications are almost endless: Adjustable-
speed drives can save 30 ~ 50% of the energy wasted in some industrial
processes; new power control systems can allow broader use of inex-
pensive ac motors in the residential, commercial, and transportation
sectors; and someday you may even have a microwave dryer for home
laundry. Power electronics provides the lever to introduce all sorts of
new electrotechnologies into the economy."

Power electronics may also help pave the way for generating electricity from renewable resources. One of the most significant potential
advances in wind turbine technology, for example, is variable-speed
generation through power electronics control. Because of shifting wind
velocities, the present generation of constant-speed turbines cannot
always deliver the maximum amount of energy and can send power
fluctuations onto utility lines. A wind turbine controlled by power electronics could provide a more optimal match to wind velocity, thus
improving energy capture by 15 ~ 20%. In addition, this next generation of wind turbines would feature simplified blade control, less physical stress on structural elements, reduced wear on drive trains, and
improved compatibility with the utility networks.

The present challenge is to reduce the cost of the power conversion
system needed for such turbines, which is still relatively expensive.
With the use of new concepts, the cost of the power electronics may be
reduced by half. To bring this about, EPRI is establishing a cooperative
research program with U.S. Windpower, Inc. (and perhaps individual
utilities) focused on development of a wind turbine controlled by power
electronics and optimized for utility use. The program would begin
with a 12-month feasibility study, followed by a development program
that would result in construction of prototype turbines. Specific goals
of the program are to clarify the role of power electronics in wind power
and to develop wind-generating units that could be fully integrated as
components of a utility system. These units could mitigate some of the
adverse electrical behavior of present wind generators and could be

automated for remote data acquisition and real-time dispatch from a central control point.

In the area of solar photovoltaics, EPRI research on power electronics should bring two major benefits. First, the low-cost, high-purity silicon materials needed for power semiconductors may also meet the requirements of high-efficiency photovoltaic cells with suitable adaptations. (Impurities or defects lower the efficiency of photovoltaic energy conversion.) Second, the power-conditioning systems that connect photovoltaic units to a utility network can be greatly improved through the use of advanced power semiconductor devices. In the past, power-conditioning systems have required addition of corrective filters containing expensive copper and iron. Advanced solid-state devices could reduce filtering requirements, as well as the physical size, weight, and cost of power-conditioning systems. At present, high-power thyristors appear to be the best candidates for constructing such systems, but by 1990, GTOs and MCTs are expected to become competitive and help further reduce the amount of control hardware.

"Many options are open to us as we develop these new systems," says Frank Goodman, Advanced Power Systems Division project manager. "The most cost-effective devices for power conditioning in solar and wind facilities have not yet been identified. The research now getting under way should help clear up the issues and eventually result in renewable energy facilities that better meet utility needs. In addition, the materials research now being aimed at improving power semiconductor devices will certainly help meet the requirements of high-efficiency photovoltaic cells as well. Advances in power electronics are definitely improving the prospects for both solar and wind energy."

Developing a national approach.   As research on improved semiconductor materials, advanced power electronics devices, and new applications has progressed, there has been a growing realization among various funding organizations that the task is too great to be handled individually, particularly if aspiring to be at the leading edge in power electronics technology. For this reason unprecedented meetings were held in Asbury Park, New Jersey, in January 1986 and then at EPRI in Palo Alto, California, in July to bring together the leaders in power electronics for exploratory talks on coordinating their research. EPRI was one of the organizers and sponsors of this workshop, which included participants from government, industry, and the university community.

Out of the workshop came tentative, informal agreement of research priorities, coordination of funding to avoid unnecessary duplication of effort, broad sharing of results, and the possibility of even closer cooperation through jointly funded programs in the future. Over the next five years, approximately $20 million will be needed for materials research and $40 million for development of improved devices. EPRI is expected to play a leading role in those programs of most use to utilities, subject to the availability of funds. Interviews with other workshop participants reveal some of the ways that sharing responsibility for various research efforts can benefit all the parties involved.

To the National Aeronautics and Space Administration (NASA), power electronics represents a critical opportunity to decrease payload weight. "Power electronics is a building-block technology for all types of spacecraft and aircraft," comments Gale Sundberg of the NASA Lewis Research Center. "We've proposed a power electronics system for an all-electric airplane, for example. Using such a system could take 10% off the empty weight of a plane and save a corresponding 10% on fuel." In such a craft, electric motors, rather than a hydraulic system, would move wing flaps and the tail assembly. Studies for a proposed space station also indicate that a power management and distribution system based on advanced power semiconductors would weigh 40% less than alternative concepts—a particularly important factor considering that thousands of dollars are required to lift each pound of equipment into orbit. Because by themselves these applications are not large enough to generate a market for particular kinds of power semiconductor devices, Sundberg emphasizes the "synergism of all our needs" with the whole spectrum of research conducted by other organizations.

Similar sentiments are echoed by Stephen Levy of the U.S. Army's Electronics Technology and Device Laboratory. "The reason it's so important to share experiences and costs is that neither the government nor industry can generate the required research money alone. This has to be a cooperative venture. All sides will profit. Most of the basic devices we are working on can ultimately be used by utilities or their major customers." As an example, Levy cites the possibility of storing energy in large superconducting coils in addition to batteries. Mechanical switching of current would be impossible in this case because of arcing, so such a storage system (useful to the army because of its light weight) would require switching by power semiconductors.

"EPRI is basically providing seed money to support development of light-fired thyristors, GTOs, and MOS-controlled thyristors," concludes Hingorani. "We cannot afford, however, to fund all the research that could eventually lead to devices of interest to the utility industry. Our priorities are very different from those of NASA and the Department of Defense, but there are clearly a large number of areas where sharing results of coordinated research could be very helpful. Again I want to emphasize the importance of leverage—although power semiconductor devices may represent a small fraction of the total cost of a utility installation, they can help save much larger fractions of total capital costs by enabling improvements in the remainder of a plant. My concern is that we are in danger of losing this leverage to other countries. R&D in power electronics requires and deserves a coordinated national approach."

10.2   The Future of Transmission—
Switching to Silicon*

> Silicon thyristor technology promises to transform the nation's transmission system with fact, "smart" solid-state switching devices that eliminate the bottlenecks encountered with conventional, electromechanical equipment.

[Refer to Figs. 10.1 and 10.2]

For the next several years, the high-voltage transmission network of the United States will face unprecedented challenges and opportunities. The challenges stem largely from a rapid increase in interutility power transfers over the network, in addition to the continued steady growth of power delivered directly from the utilities' power plants to their customers. An increasing number of cogenerators and independent power producers are also demanding access to the network. In the United States, as in many other parts of the world, transmission systems were not designed to cope with the present magnitude of bulk power transfers and third-party access. Moreover, the construction of new lines to handle load growth and increased numbers of bulk transactions has been stymied for a variety of regulatory, environmental, and public policy reasons. As a result, America's transmission network now has many limitations, which represent a difficult challenge for the future.

There is a new opportunity to handle the immediate challenge, while establishing an improved foundation for long-term growth, through the development of technology that can make the present transmission system more flexible. "In 20 years the U.S. transmission system will look very different from the present system," says Narain Hingorani, EPRI vice president, Electrical Systems Division. "Solid-state, high-power electronic switches will provide much greater control speed and flexibility. The whole system will also become more highly automated, enabling dispatchers to direct more power over selected lines."

Unlike the switched networks used in telephone systems to direct calls to a specific party, alternating-current (ac) electric power transmission networks are not easily controlled in terms of directing the

---

* This article was written by John Douglas, science writer. Technical background information was provided by Narain Hingorani, Frank Young, and Robert Iveson, Electrical Systems Division.

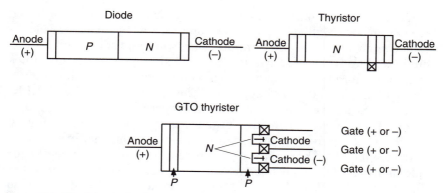

Figure 10.1 Thyristor technology.

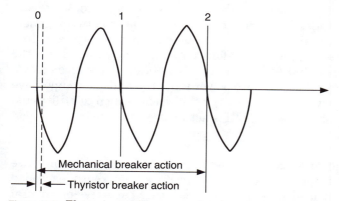

Figure 10.2 Eletronic speed in a power breaker.

flow of power. Because today's utility transmission systems are so closely interconnected, this lack of control means that a bulk transfer of power from Ontario Hydro power plants to New York City, for example, may not follow the most direct corridor of lines and may interfere with transmission operations as far away as Ohio. This phenomenon is known as *loop flow*.

EPRI has developed a vision of the advanced transmission system utilities need to cope with increasingly demanding conditions. Called FACTS (for Flexible Ac Transmission System), this system could overcome loop flows and provide a more sophisticated degree of control through a combination of high-power electronic devices and computerized system automation. Specifically, large solid-state switches, or thyristors, would be used in various types of controllers to manipulate alternating current on transmission lines. With such thyristor-based devices and with high-speed coordination provided by an advanced system of communications and computer hardware and software, utilities would be able to increase the loading on existing transmission lines and handle more transactions without compromising reliability.

Recognizing both the challenges and the opportunities ahead, EPRI has launched a seven-year, $20 million research program devoted to FACTS. This effort will involve numerous specific projects, ranging from the evaluation of existing transmission system performance to the development of hardware and the coordination of strategies for implementing FACTS on host utility systems. Some thyristor-based devices related to FACTS have already been developed and tested, so utilities have begun to incorporate particular elements of FACTS to meet the needs of their own transmission systems.

"We can no longer assume that the electric power system of today represents mature technology," Hingorani says. "In the long run, power will be controlled by thyristors at several stages between generation and end use. Utilities will be able to manage their systems to a much greater extent than ever before possible. But research on the technologies involved in FACTS is urgently needed. Transmission systems are the lifeline of the electric power industry, and the competitive position of electricity among the energy choices now open to utility customers is at stake. That's why EPRI is acting decisively to take the lead in this vital area."

Most complex machine. According to a recent EPRI study, the increase in bulk power transfers over transmission lines has greatly exceeded the growth rate of load and that of generating capacity for several years. This increase includes both the interchange of power between utility members of a power pool and the wheeling of power across a utility's lines, even though the utility may not be involved as either buyer or seller. Taken together, such bulk transactions now amount to roughly one-third of electricity sales to ultimate customers.

Not surprisingly, given the limited construction of new lines, bottlenecks are appearing in areas with overloaded transmission systems. Voltage and frequency fluctuations, in particular, are becoming more common as free flows of power spread across interconnected systems. Most of the devices available today for taming wayward loop flows or stabilizing voltages are mechanically switched—meaning that system response to changing conditions is slow and device maintenance costs are high. There is an immediate need for faster, more reliable alternative control technologies.

"The U.S. power system is perhaps the most complex machine ever created by man," asserts Frank Young, director of the Electrical Systems Division. "But we're asking this machine to perform functions that weren't even conceived of by its original designers. In some cases, a transmission line can be upgraded by raising voltage or adding new conductors along an existing right-of-way. A more fundamental solution, however, is to use high-power electronic technology to make the whole system 'smarter' and thus more responsive to changing needs."

Young emphasizes that the transition from mechanical to electronic control will require several stages of development and implementation. The first step—production of semiconductor materials capable of handling high voltage and current—has already largely been accomplished (*EPRI Journal,* December 1986). Next comes the step of designing and testing a new generation of solid-state devices, generally based on thyristor technology, that can perform the control functions needed by transmission systems. Finally, these materials and devices—the ingredients for a "second electronics revolution"—must be demonstrated in actual utility system operation. In addition, a parallel effort must be undertaken to develop communication and computer capabilities that can take advantage of the greater speed and flexibility provided by the new control hardware.

The bottom line of this FACTS development effort can be measured both physically and economically. The ultimate physical capacity of a transmission system is set by the so-called thermal limit of its lines— roughly speaking, the amount of power that can be carried before conductors start to overheat and sag toward objects below. Portions of today's transmission systems are restricted to power levels well below their ultimate limit because of other factors, such as voltage and stability. In some cases, introducing FACTS technology could double the amount of power carried over a line. Taking even a conservative case of a 25% power increase on a 2000-MW line, the economic benefit to the utility would be approximately $10 million per year.

Three basic approaches. Increasing the power transfer capability of a line usually involves one of three basic approaches: maintaining the proper voltage, optimizing the power angle (phase difference between voltages at opposite ends of a line), or changing the impedance (total opposition to ac flow). Each of these parameters can be controlled at high speed by thyristor-based devices now being developed or tested as part of EPRI's FACTS program.

Controlling voltage in an ac circuit is complicated by the existence of reactive power (expressed as volt-amperes reactive, or VARs). Reactive power is the power required to maintain the electric and magnetic fields surrounding a power line. Electric fields tend to raise the voltage on a lightly loaded line. Their effect is counteracted by connecting shunt reactors onto the line. Magnetic fields associated with heavy current flow tend to reduce the voltage on a line. That effect is counteracted by connecting shunt capacitors.

On some power systems, it is possible to experience both extremes of voltage in a single 24-hour period. To handle such cases, shunt reactors and shunt capacitors have been combined with a thyristor-controlled switch into a device called a static VAR compensator, or SVC. Thanks to the thyristor switch, an SVC can operate not only to make steady-state voltage corrections but also to respond very rapidly during a disturbance on the line—minimizing voltage depression during a fault and voltage overshoot during fault recovery. Some SVCs are already commercially available and will play a major role in FACTS voltage control strategies. Research to improve the design of SVCs, so that their size and cost can be greatly reduced, is also under way.

Power angle changes can sometimes be used to solve the problem of loop flows. A few regions now use phase-shifting transformers to redirect power flow from one line to another. Changing the power angle setting of such a transformer, however, which involves a motor-driven tap, can take as long as a minute and a half. Also, the number of resettings per day is usually limited to 10 or so, since tap changes create arcing and thus eventually wear out the equipment. Using thyristors to change the tap setting would reduce the time involved to milliseconds and, because there are no moving parts, would virtually eliminate the need to restrict the number of daily resettings. Developing such an advanced power angle regulator is one of the specific goals of the present research program.

Devices for rapidly changing the impedance of a transmission circuit—a capability previously unavailable—will prove useful in a variety of ways. One such device is currently undergoing utility demonstration; three others are to be developed as part of the FACTS program. Each is designed to counteract a specific problem.

Southern California Edison, for example, has in commercial use a thyristor-controlled device that suppresses low-frequency oscillations which can occur when large capacitor banks are inserted in series with the circuit to reduce impedance on long transmission lines. Called the NGH subsynchronous resonance damper after its inventor, EPRI's Narain G. Hingorani, the device automatically changes the impedance of a transmission circuit enough to keep its voltage from resonating at frequencies below 60 Hz. Earlier, such resonance problems had become severe enough at one Southern California Edison plant to damage a generator shaft by creating a heavy torque on it.

Another approach to changing the impedance of a transmission circuit is to have a thyristor directly control a variable amount of capacitance inserted in series on a line. Current practice involves adding capacitors to a line by switching them in large, discrete units. A modular series capacitor with thyristor control would be able to dial in or out just the right amount of capacitance during steady-state conditions and could also react quickly enough to help stabilize the transmission system during a disturbance.

One extreme type of disturbance, the voltage surge caused by a lightning strike on a line, requires an additional measure of protection. Surge arresters give this protection by providing a low-impedance path to ground when voltage rises above a specified level. The current generation of metal oxide surge arresters depends on the properties of certain materials (such as zinc oxide) to reduce resistance sharply as voltage increases. The threshold voltage for such devices, however, cannot be set precisely or changed easily. By using thyristor control to ground a voltage surge, advanced arresters will have increased precision, flexibility, and safety, and probably a lower cost.

The final type of impedance-changing device now under development addresses the problem of how to dissipate energy quickly when a generator is suddenly cut off from part or all of its load—for example, during a line fault nearby. If no attempt is made to absorb energy from the generator, it can quickly begin to speed up and force a plant shutdown lasting hours. If a large resistor installed at a power plant could be brought on-line when needed to convert the generator's energy to heat, the unit could continue to operate at a steady speed and then be reconnected promptly when the fault was cleared. Such resistors have found only limited use so far because they are not variable and the mechanical switches that control them are slow. A dynamic load brake controlled by thyristors would be able to add just the right amount of resistance to reduce the acceleration of a generator and allow it to remain synchronized with the system.

"The basic methods and materials used to control transmission systems haven't changed substantially over the last 25 years," says Robert Iveson, technical adviser in the Electrical Systems Division. "FACTS will change that. Static VAR compensators and NGH dampers are already in utility use. Power angle regulators, modular series capacitors, high-energy arresters, and dynamic load brakes with high-speed controllers are being developed as part of our current FACTS research. The system is already beginning to evolve—and the end result will be a real revolution in power transmission."

DC and other application. One of the first applications of thyristor technology was the conversion of ac to dc power, and the use of dc lines embedded in the larger ac network can be an important part of FACTS. Such lines are ideal for delivering bulk power across large distances between specific locations by effectively reducing the power angle difference between them. The 800-mile Pacific Intertie, for example, already provides a dc link between the hydroelectric plants of the Pacific Northwest and the rapidly growing cities of the Southwest. The primary advantage of such dc ties is that they add greater control and stability to the total transmission system. Power flow can be directed at will from point to point.

The growing capabilities and declining cost of thyristors may enable them to be used in a variety of power control devices beyond those currently under development. Modular series reactors, which would employ thyristors to add variable amounts of inductance to a line, could be used to limit the flow of power through underground cables to avoid overloading them. Thyristor-controlled ferroresonance dampers could prevent the oscillations that sometimes occur in large transformers at the end of a long power line—causing them to explode. In a fault current limiter, thyristors would quickly change the impedance of a faulted line to prevent currents from growing large enough to cause equipment damage.

The ultimate use of thyristors would be as circuit breakers. This application would be particularly important in regions with frequent power interruptions due to storms, where reduced maintenance costs could make these breakers competitive with the more common, mechanical types. In addition, faults could be cleared almost instantaneously, minimizing damage to equipment and interruption to customers.

The timetable for investigating these and other potential applications depends largely on how quickly thyristor technology itself improves. The power-handling capability of a thyristor depends in part on the diameter of the silicon wafer from which it is fabricated. Today's largest commercial thyristors are made from 100-mm (4-in) wafers, but recent advances in materials research have brought a 150-mm (6-in) chip close to commercialization. New package designs, which replace heavy, expensive metal contact surfaces with silicon, will greatly reduce the cost the size of future thyristors.

At the same time, the internal structure of thyristors is also changing. The greatest barrier to more widespread use of thyristors in transmission applications has traditionally been that, although easy to turn on, they are difficult to turn off. Now a new gate-turnoff (GTO) thyristor—invented in the United States and recently commercialized in Japan—offers a partial solution. The GTO thyristor may simplify the circuitry of some devices, such as the power conditioning equipment being used at Southern California Edison's new battery energy storage facility. However, the higher losses and cost of GTOs, compared with conventional thyristors, will probably continue to limit their use.

A more promising technology, currently being developed by EPRI, involves turning a thyristor off and on by means of metal oxide semiconductor (MOS) microcircuitry placed directly onto the surface of the thyristor wafer. This MOS-controlled thyristor, or MCT, would be able to switch large line currents by using a very small control current. Its turnoff speed is less than one-third of the GTOs, and it has considerably lower energy losses. MCTs are expected to be available for use in ac-dc conversion and other transmission applications within three to five years.

Open to ideas. EPRI's research program to develop the many aspects of FACTS began in late 1988 and is divided into two phases. The first phase, which is expected to last three or four years, will involve the evaluation of the technical and economic benefits of FACTS and the development of design specifications for specific power control devices. The thermal performance of present transmission systems will also be evaluated and overall control strategies worked out. The second phase, involving actual hardware development and demonstration on utility systems, could begin as early as 1990.

"We need to follow a systems approach to FACTS, even at the earliest stages," Frank Young emphasizes. "Although it can be implemented piecemeal, the pieces have to fit together. By the time prototypes of phase angle transformers and modular series capacitors are ready for utility demonstration, the strategies for integrating them into a transmission system must also be available. Just as existing software helps an operator dispatch generators, for example, FACTS-related software will help him dispatch impedance, phase angle, and voltage as well."

Economic and social forces, in addition to the technological developments discussed here, will help shape transmission systems of the future. Increased competition in the utility industry, for example, may bring fundamental changes to the U.S. transmission network, such as the development of a new hierarchy of control centers and the "unbundling" of services according to different customer needs. And the addition of generating capacity in smaller increments, more widely dispersed geographically, may contribute to the difficulty of maintaining control of transmission systems. Such changes are likely to create a demand for even more flexibility and control, requiring still further technological advances.

"We need to keep looking for new ways to utilize thyristor technology to satisfy changing needs," concludes Narain Hingorani. "Think of how many different uses have been found for transistors. Right now we're working on only half a dozen or so configurations of thyristors, and we must remain open to new ideas. I particularly want to see more people get involved—for example, those now working on end-use technology—so that we can find more ways to adapt the FACTS concept to meet the requirements of utilities and their customers."

Supplemental Special Problems

1   Show why electromagnetic rather than electrostatic machines are universally used for power generation. Consider a value of $B = 10^4$ lines/cm^2 and electric field strength $E = 3 \times 10^4$ volts/cm at atmospheric pressure.

2   In a static induction machine, show the significance and effects of the magnetizing reactance on the machine parameters, and what approximations can be done in this equivalent circuit. Comment if the core is a nonmagnetic material.

3   In the circuit shown find the current i where

$$\frac{dB}{dt} = 10^8 \text{ lines/cm}^2\text{/sec}$$

Area of the loop is a square.

4   Explain the following:
   a. A coil linking a magnetic flux suddenly is short-circuited.
   b. A magnetic circuit is subjected to a deformation resulting in an air gap.
   c. Increasing the DC bias of a saturable-reactor.
   d. Decreasing the AC current in a saturable-reactor.
   e. Increasing the AC current beyond the maximum impedance value in a saturable-resistor.

5   Prove in detail that in a static induction machine the voltage across a load resistance will be a little less than 70 percent of the no-load voltage (with constant voltage on the primary).

6   *a.* Sketch a straight line approximation for the impedance characteristic against line current of a single-phase saturable-resistor, and then express such characteristic mathematically.
   b. Express in mathematical terms the effect of injecting DC bias in a central leg of a two-window saturable-resistor.

7 In reference to Kelvin's law where, in a singly excited magnetic circuit without saturation, a deformation takes place at constant current, half of the supplied energy goes into the action of deformation while the other half is magnetically stored. Using the concept of stored magnetic energy density as $B^2/2\mu$ joules/m³, obtain an expression for the mechanical work done in terms of the magnetic circuit elements.

8 ALNICO is considered a hard magnetic material. Under the incidence of sinusoidal magnetic intensity of ampere-turns/m, secure an expression for the induced electric field at some depth inside the magnet.

9 Extend Prob. 8 to obtain expressions for the surface impedance and the rate of propagation of the magnetization plane. Assume the ALNICO material is thick.

10 Using the concept of the constant linkage theorem, show that the closed circuit current flow will remain constant indefinitely if the circuit resistance is zero. Predict the kind of circuit material which can possess zero resistance.

11 Establish an expression for torque in induction machines using the law of virtual displacement if the resultant air-gap flux density is expressed by

$$B = B_1 \sin P\theta - B_2 \sin(\theta + \alpha)3$$

Let

$$k_{wm} = k_{dm}k_{pm}$$

where P = Number of poles
k_{wm} = the winding factor for mth harmonic
k_{dm} and k_{pm} = the distribution and pitch factor, respectively

Obtain the order of m for optimum k_{wm}

12 Applying the rules of symmetrical components, establish the mode of interconnection among the three sequence networks for a line-ground fault in terms of the machine leakage reactances and magnetizing reactance for induction motor at no-load.

13 Repeat Prob. 12 for a line-line short circuit for an unloaded induction motor.

14 Repeat Prob. 12 for a double-line-to-ground short circuit for an unloaded induction motor.

15 Show that the net torque of a polyphase induction motor is the result of a dominant forward field component with respect to a much smaller backward field component.

16 Using an approximate equivalent circuit for the polyphase induction motor (by neglecting) the stator impedance, derive the condition on the slip (s) for maximum electromagnetic torque.

17 Derive the exact equivalent circuit for the polyphase induction motor in systematic steps starting from that of a lossy transformer.

18 Using the exact equivalent circuit of the polyphase induction motor, establish the condition on the slip (s) for maximum electromagnetic torque. S is the independent variable.

19 Repeat Prob. 18, but for maximum output mechanical torque.

20 Repeat Prob. 18, but for maximum output efficiency.

21 Using the exact equivalent circuit of the polyphase induction motor, establish the ratio of starting torque to starting line current.

22 From the solution of starting torque to starting current ratio established in Prob. 21, find the condition on s for the maximum of the ratio.

23 With the insertion of a saturable-resistor in series with the rotor in a three-phase induction motor, establish the ratio of starting torque to starting current. Compare the new ratio to that without saturable-resistor.

24 Calculate the starting line current of the polyphase induction with and then without saturable-resistor in the rotor, using the exact equivalent circuit.

25 From Prob. 24, establish the condition on the slip (s) for maximum current at an optimum efficiency.

26 Using the exact equivalent circuit of the polyphase induction motor, establish condition on the slip (s) for optimum overall power factor of the motor circuit.

27 For a 1000-hp induction motor, calculate the ventilation loss in kw if the velocity of delivered air is 5000 ft/min.

28 Establish an exact equivalent circuit for the high-impedance rotor induction motor, and then calculate the condition on the slip (s) for maximum electromagnetic torque.

29 Repeat Prob. 28 for the deep-bar rotor induction motor.

30 Repeat Prob. 28 for the idle-bar rotor induction motor.

31 Repeat Prob. 28 for the double squirrel-cage rotor induction motor.

32 For the series hysteresis rotor, establish the ratio of electromagnetic torque with respect to the total line current. Then identify condition on s for the maximum ratio.

33 With the insertion of saturable-resistor in a series connection of the deep-bar rotor induction motor, establish the ratio of the total mechanical output torque with respect to line current.

34 From Prob. 33, establish the operational condition on the slip (s) for maximum ratio of output mechanical torque with respect to line current.

35 Repeat Prob. 33 for the double squirrel-cage rotor induction motor.

36 Repeat Prob. 34 for the idle-bar rotor induction motor.

37 For a polyphase induction motor with idle-bar rotor, obtain expressions for the electromagnetic torque and line current at low and high drive, and then identify the condition of maximum torque.

38 *a.* Show mathematically the significance of using ferromagnetic core materials in magnetic circuits.
    b. For a quasi-static magnetic field system, write the integral form of Maxwell field equations, and then the terminal voltage for a system with N electrical terminals pairs and M mechanical terminal pairs (one-dimensional displacement).

39 Establish an equivalent circuit of a polyphase induction motor with:
    a. Saturable-reactor connected in series with the rotor circuit.
    b. Repeat part *a* with a saturable-resistor in the rotor.

40 Consider the effect of DC bias on both the saturable-reactor and saturable-resistor; discuss the use of these elements on induction motor equivalent circuit, especially on the ratio of torque/amp.

41 Consider an AC machine with a three-phase winding on the stator where N_s is the total number of turns on each stator phase and N_r is the total number of turns in the rotor winding. The surface current densities produced by the three armature currents on the surface at $R + g$ are:

$$k_a = i_z \frac{N_s i_a}{2(R + g)} \sin \psi$$

$$k_b = i_z \frac{N_s i_b}{2(R + g)} \sin \left(\psi - \frac{2\pi}{3}\right)$$

$$k_c = i_z \frac{N_s i_c}{2(R + g)} \sin \left(\psi - \frac{4\pi}{3}\right)$$

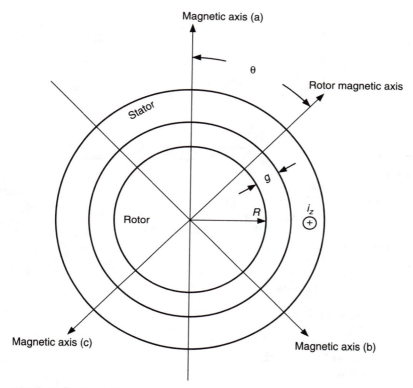

Magnetic axis (a)

θ

Rotor magnetic axis

Stator

g

R

i_z

Rotor

Magnetic axis (c)

Magnetic axis (b)

Figure A.1 Problem 41

The surface current density due to the rotor current on the surface at R is

$$k_r = i_z \frac{N_r i_r}{2R} \; sin(\psi - \theta)$$

Assume $g \ll R$ so that there is no appreciable variation in the radial component of magnetic field across the air gap (see Fig. A.1).

    a. Find the radial flux density due to current in each winding.

    b. Write the electrical terminal relations for the machine.

    c. Find an expression for T_{em}.

42 The system shown in Fig. A.2 is composed of a pair of parallel plates of width W and separation d, connected by a sliding conductor of mass M. The sliding conductor makes frictionless and perfect electrical contact with the plates. The entire system is immersed in a static magnetic field H_0 into the page, and the plates are excited by the battery V_0 through the resistance R and switch S. All conductors may be assumed to be perfect; fringing fields may be neglected; and you may assume the $i/W \ll H_0$.

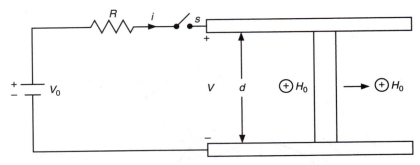

Figure A.2 Problem 42

   a. Find the force on the sliding conductor in terms of current i.
   b. With the system at rest, the switch S is closed at $t = 0$. Find the velocity of the sliding conductor $v(t)$ for $t > 0$.

43 Show that a sinusoidal flux distribution in the air gap of a cylindrical rotor AC machine requires a sinusoidal current distribution of:

$$J = \frac{1}{r} \frac{p}{2} \frac{Bm}{\mu_0} g \sin \frac{p}{S} \theta_s$$

J = linear current density in amp/meter
r = mean radius of the air gap
P = number of poles
B_m = maximum flux density
g = air-gap width
s = angular location of a point in the air gap as a reference

44 Establish the variable frequency equivalent circuit for a polyphase wound rotor induction motor.

45 Establish the solution for the total permeance wave equation by summarizing the contribution from the stator and slot openings, and then by their product. Identify the operational advantage of each solution.

46 Calculate the sound intensity at a point of 54″ from the centerline of a polyphase induction motor having the following data:

   $B = 10^{-3}$ web/in^2
   D = air-gap diameter = 30″
   P = number of poles = 6
   f = frequency of the impressed field = 60 Hz
   the machine rating = 20 hp
   the stator outside diameter = 40″

Consider only the effect of the fundamental field.

47 *a.* Write the basic effects of harmonic fields present with the fundamental field on transient and steady-state operations of rotating machines.

  b. Referring to asynchronous crawling phenomenon of an induction machine, examine the prospects of obtaining synchronous torque as steady supply source.

48 Establish the equivalent circuit of a polyphase hysteresis type induction motor and then calculate: torque, line current, and rotor power factor (squirrel-cage rotor of hysteresis material). Then find the condition of maximum torque.

49 *a.* For a two-pole, three-phase winding:

$$i_a = I_a sin(wt - \theta)$$

$$i_b = I_b sin\left(wt - \theta - \frac{2\pi}{3}\right)$$

$$i_c = I_c sin\left(wt - \theta - \frac{2\pi}{3}\right)$$

Calculate the net mmf at $t = 0$

  b. From the equivalent circuit of a three-phase induction motor with a saturable-reactor in the rotor, find the condition of maximum torque and then express the value of that torque.

50 *a.* Compare the electric and magnetic operational properties between a saturable-reactor and a saturable-resistor.

  b. Establish a ratio of torque/amp for a polyphase induction motor; then indicate the effects of inserting a saturable-reactor in the rotor.

51 Using the generalized equivalent rotating field circuit for a two-pole machine, calculate the pulsating double frequency torque.

52 Consider an asymmetrical components equivalent circuit of a single-phase induction motor.

  a. What criticism can you direct to such an equivalent circuit? Explain it clearly and indicate its significance.

  b. Why is the zero-sequence component not represented?

53 Refer to Prob. 52:

  a. Indicate and establish the source of mechanical vibrations.

  b. Express the pulsating torque of the forward and backward fields.

54 Refer to Prob. 52:

  a. Illustrate by means of a simple single-phase AC circuit the presence of the double-frequency power.

  b. Why does this machine have zero torque at starting?

  c. Compare the double-revolving field theory with respect to the cross-field theory in determining the machine performance.

d. Compare the performance of the single-phase motor with a polyphase machine running with one phase disconnected from the supply.

55 Derive the equation of voltage for a machine having a salient pole rotor without winding. Consider it an eddy-current machine.

56 Derive the torque equation for a salient pole machine whose inductance is:

$$L = \frac{1}{2}(L_d + L_g \cos\,\theta) + \frac{1}{2}(L_d - L_g \sin\,\theta)\cos\,2\theta + \Delta L \cos\,S\theta_2$$

where *s* is variable.

57 Transform the equation of motion from a holonomic to a nonholonomic system.

58 *a.* Give analytical descriptions for the following:
  1. Affine connection
  2. Magnetic field tensor
  3. The G tensor
  4. The V tensor
  5. Christoffel symbol
  b. Write the voltage equation for a machine in the quasi-holonomic frame. Assume the rotor has no winding.

59 There are four references used to represent modeling of electromechanical devices. Write the tensor torque equation in each frame, and extract the corresponding voltage equations.

60 *a.* Write a general form for the dynamic equation of rotating machines in terms of the absolute time derivative. Comment on the significance of each term.
  b. What is meant by the absolute frequency tensor? Write the mathematical form and comment on each term.
  c. Write Ohm's law and torque equations in terms of the absolute frequency tensor.

61 Refer to the variable frequency equivalent circuit of synchronous machines. Write expressions for the undirectional and then for the alternating torques.

62 Voltage surge in the form of $v(t) = e(x,t)sin\,wt$ is suddenly subjected on a three-phase wound rotor induction motor having saturable-resistor inserted into its rotor. $e(x,t)$ is the voltage induced across a capacitor shunting a transmission line just at the entry to the motor, generated by a sustained lightning surge. Obtain the solution for the time response of current induced into the rotor circuit at a very small slip. Use the expression for $e(x,t)$ given in this text.

63   Repeat Prob. 62, but for a slip corresponding to low drive, i.e., a relatively large slip.

64   From the solution of time response of rotor current obtained in Prob. 63, obtain an expression for the electromagnetic torque at $t = 0.2$ sec and slip $s = 0.10$.

65   Repeat Prob. 62 if $e(x,t)$ is the voltage induced on a transmission line having a series inserted ideal inductor located at the entry of the motor.

66   Repeat Prob. 65, but for a large slip corresponding to a low drive operation.

67   From the solution of the time response of rotor current obtained in Prob. 66, secure an expression for the electromagnetic torque at $t = 0.2$ sec and $s = 0.10$.

68   Voltage surge in the form of $v(t) = e(x,t) + V_0 U_{-1}(t)$ is subjected to a three-phase synchronous motor having saturable-resistor inserted in its stator. $e(x,t)$ is the voltage induced across an ideal capacitor shunting a transmission line just at the entry to the three-phase motor, due to an actual lightning surge. Obtain the solution for the stator current at any time (t). The motor was operating at full load in a lagging power-factor condition.

69   From solution of the time response of stator current obtained in Prob. 68, identify the new status of the synchronous motor power factor.

70   A power system substation containing a three-phase transformer connected to a three-phase induction motor operating at full load is subjected suddenly to current surge induced by actual lightning surge. The incident current surge is the output of ideal inductor where the input is of the form: $i(t) = I(x,t) + I_0$. I_0 is initially stored current in the inductor at $t = O^+$. Obtain the solution for the current induced into the induction motor rotor as a function of time.

71   From the solution of the time response of rotor current obtained in Prob. 70, secure the solution for the electromagnetic torque at any slip.

72   An actual lightning voltage surge is suddenly incident on a transmission line just before a series ideal inductor followed by a shunt capacitor. The L-C network is connected to a three-phase transformer. Obtain the solution for the current induced in the transformer at any time (t). Assume the transformer ideal.

73   From solution of the time response of surge current in the transformer secured in Prob. 72, establish an expression for the induced magnetic field intensity vector $\overline{H}$ as a function of time. Also obtain the solution for the induced electric field intensity vector E as a function of time. To obtain solutions for $\overline{H}$ and $\overline{E}$, you to may resort to Maxwell's field equations.

74 An actual lightning current surge is incident at the entry of a saturable-resistor shunt capacitor installed at a transmission line feeding a three-phase alternator operating at full load. The alternator is of cylindrical armature. Obtain the solution for the current induced into the armature as a function of time.

75 Repeat Prob. 74 if the armature is of salient poles rotor having P poles.

76 Solve Prob. 75 if the protective network involves a shunt capacitor followed by a series saturable-resistor.

77 Consider a three-phase induction motor of wound rotor type having saturable-resistor inserted in its rotor. The motor is fed from supply lines at which a three-phase saturistor intervenes just at the motor terminals. A rectangular voltage pulse of the form: $v(t) = AU_{-1} - AU_{-1}(t - T)$ is suddenly incident at the supply lines. Obtain an expression for the time response of rotor current while the motor was running at full-load slip.

78 From results obtained in Prob. 77, secure an expression for the electromagnetic torque, and then identify the condition on the slip (s) for maximum torque.

79 Repeat Prob. 77 if an ideal inductor is connected in series with the saturable-resistor at the feeding lines.

80 A protective network comprised of shunt capacitor followed by a series saturable-resistor is connected at the supply lines preceding a three-phase transformer to a three-phase synchronous motor. An impulse voltage surge is incident at a point just before the capacitor. Obtain the solution for the current-time response in the synchronous motor stator. Assume the rotor has cylindrical armature.

81 In reference to Prob. 80, obtain an expression for the overall system power factor.

82 A three-phase power system is fed from two generating sources and operating at full load. Total line impedance of a line connecting the two sources is $100\ \Omega$. The source at the left has an internal impedance of $10\angle 15°\ \Omega$. At $t = 0$, one phase is suddenly disconnected and a short circuit occurred between the remaining phase. Calculate all the symmetrical components of currents and voltages. Location of faults is at a line impedance of $25\ \Omega$ from the left source.

83 From the information and solution of sequence symmetrical components established in Prob. 82, obtain graphically the span of phase angle protection coverage for operational distance relay.

84 Repeat Prob. 82 if a double-line-to-ground short circuit occurred for the remaining two.

85 From the information and solution of sequence symmetrical components obtained in Prob. 84, establish graphically the span of phase angle protection coverage for operational distance relay.

Index

ABOUT THE AUTHOR

Khalil Denno is Distinguished Professor in the Department of Electrical and Computer Engineering at the New Jersey Institute of Technology in Newark. He is also the author of *High Voltage Engineering in Power Systems, Engineering Economics of Alternative Energy Sources,* and *Power System Design and Applications of Alternative Energy Sources.* Professor Denno is a Fellow of the IEE (Great Britain) and a Senior Member of the IEEE.

DATE DUE

REC'D NOV 2 8 2001			